Physiological Genetics

PHYSIOLOGICAL ECOLOGY

A Series of Monographs, Texts, and Treatises

EDITED BY

T. T. KOZLOWSKI

University of Wisconsin
Madison, Wisconsin

T. T. KOZLOWSKI. Growth and Development of Trees, Volumes I and II — 1971

DANIEL HILLEL. Soil and Water: Physical Principles and Processes, 1971

J. LEVITT. Responses of Plants to Environmental Stresses, 1972

V. B. YOUNGNER AND C. M. MCKELL (Eds.). The Biology and Utilization of Grasses, 1972

T. T. KOZLOWSKI (Ed.). Seed Biology, Volumes I, II, and III — 1972

YOAV WAISEL. Biology of Halophytes, 1972

G. C. MARKS AND T. T. KOZLOWSKI (Eds.). Ectomycorrhizae: Their Ecology and Physiology, 1973

T. T. KOZLOWSKI (Ed.). Shedding of Plant Parts, 1973

ELROY L. RICE. Allelopathy, 1974

T. T. KOZLOWSKI AND C. E. AHLGREN (Eds.). Fire and Ecosystems, 1974

J. BRIAN MUDD AND T. T. KOZLOWSKI (Eds.). Responses of Plants to Air Pollution, 1975

REXFORD DAUBENMIRE. Plant Geography, 1978

JOHN G. SCANDALIOS (Ed.), Physiological Genetics, 1979

In Preparation

J. LEVITT. Responses of Plants to Environmental Stress, 2nd Edition. Volume I: Chilling, Freezing, and High Temperature Stresses, 1979, Volume II: Water, Radiation, Salt, and Miscellaneous Stresses, 1980

BERTRAM G. MURRAY, JR. Population Dynamics: Alternative Models, 1979

JAMES A. LARSEN. The Boreal Ecosystem, 1980

Physiological Genetics

EDITED BY

JOHN G. SCANDALIOS

Department of Genetics
North Carolina State University
Raleigh, North Carolina

1979

ACADEMIC PRESS
A Subsidiary of Harcourt Brace Jovanovich, Publishers
New York London Toronto Sydney San Francisco

ACADEMIC PRESS, INC.
111 Fifth Avenue, New York, New York 10003

United Kingdom Edition published by
ACADEMIC PRESS, INC. (LONDON) LTD.
24/28 Oval Road, London NW1 7DX

Library of Congress Cataloging in Publication Data

Main entry under title:

Physiological genetics.

   (Physiological ecology series)
   Includes bibliographies and index.
   1. Genetics. 2. Physiology. I. Scandalios, John G.
QH430.P49   575.2'1   79–50219
ISBN 0–12–620980–4

PRINTED IN THE UNITED STATES OF AMERICA

79 80 81 82   9 8 7 6 5 4 3 2 1

Contents

v

4   Biochemical–Genetic Control of Morphogenesis
William A. Scott

5   Molecular Bases of Cytoplasmic Male Sterility in Maize
C. S. Levings III and D. R. Pring

6   Somatic Cell Genetics of Higher Plants: Appraising the Application of Bacterial Systems to Higher Plant Cells Cultured *in Vitro*
Donna Parke and Peter S. Carlson

List of Contributors

Numbers in parentheses indicate the pages on which the authors' contributions begin.

Peter S. Carlson (195), Department of Crop and Soil Sciences, Michigan State University, East Lansing, Michigan 48824

Tuan-Hua David Ho (109), Department of Biology, Massachusetts Institute of Technology, Cambridge, Massachusetts 02139, and Department of Botany, University of Illinois, Urbana, Illinois 81801

George Johnson (239), Department of Biology, Washington University, St. Louis, Missouri 63130

C. S. Levings III (171), Department of Genetics, North Carolina State University, Raleigh, North Carolina 27650

K. Paigen (1), Department of Molecular Biology, Roswell Park Memorial Institute, Buffalo, New York 14263

Donna Parke (195), Department of Crop and Soil Sciences, Michigan State University, East Lansing, Michigan 48824

D. R. Pring (171), Department of Plant Pathology, University of Florida, Gainesville, Florida 32611

John G. Scandalios (63), Department of Genetics, North Carolina State University, Raleigh, North Carolina 27650

William A. Scott (141), The Rockefeller University, New York, New York 10021

Preface

Physiological genetics is the study of gene action at all levels and encompasses the genic control of both metabolism and development. No sharp distinction can be drawn between these two broad aspects of experimental biology. However, although both concern ontogeny, they deal with it at different levels. Studies of metabolic systems involving different genotypes usually involve specific metabolic pathways, genetically dissecting individual biochemical reactions, and determining how these are affected by specific enzymes or sets of enzymes encoded in identifiable genes. Developmental studies often involve the identification of specific developmental differences, both temporal and spatial, and subsequently attempting to identify, usually by working backwards, the genetic differences that lead to such developmental diversification. Thus, physiological genetics is that sub-branch of the science of genetics dealing with the study of developmental events or processes through which genetic differences are eventually expressed as phenotypic differences. It is essential, therefore, that a thorough understanding of phenotypic differences involve an understanding of how such differences are generated both spatially and temporally, and how structural genes may respond to both external and internal signals to be so precisely expressed in time and place.

Until relatively recently the primary concern of genetics was to define the gene, its structure, function, mutation, and replication. This was achieved by the progressive "molecularization" of genetics. It was correctly assumed by most geneticists that it would have been virtually fruitless then to attempt to understand the intervention and sequential action of genes in the realization of the phenotype, especially when considering the complexity of higher organisms. Nevertheless, a number of early studies, with both plants and animals, combining genetics, experimental embryology, and physiology (e.g., pigment production in some animals and plants, and metabolic pathways in microbes) proved most constructive in efforts to understand gene action at least at the phenomenological level.

The basic problem of how genes function in the complex higher eukaryotes has only just recently come to the forefront of modern experimental biology. This is largely due to new and highly sophisticated knowledge and technology in the broad area of molecular biology which allows us to experimentally probe some old questions so that new ones can be generated and, hopefully, answered. Information generated as to how genes operate in lower organisms has suggested new and exciting approaches to problems in the higher and more complex forms. Differential gene function, responsible for generating a variety of cell phenotypes from a single genotype, has become the central problem of contemporary experimental biology.

In a rapidly developing field of science the ability to keep informed of progress becomes a major problem. Consequently, there is an occasional need to summarize and highlight some of the more interesting and, it is hoped, significant advances as a useful reference source for our interested colleagues. Additionally, the material should also prove useful for teaching and training at the graduate and advanced undergraduate levels. Only time will render a verdict on the success of these goals.

The chapters were not intended as comprehensive reviews, but represent personal accounts of developments best known to each writer who, in each case, was involved in the research that has led to significant advances in the respective areas. Each contributor provides a current and critical account of recent advances and a number of exceptional model systems for studies in physiological genetics. The book covers this area from several different points of view. Following an introductory chapter describing genetic factors in developmental gene regulation, there are discussions on enzyme differentiation, hormonal control of gene expression, biochemical genetics of morphogenesis, cytoplasmic male sterility in maize, plant somatic cell genetics, and the population dynamics of genetic polymorphism. We hope we have covered some material of lasting value in view of the rapid developments in this field.

I am indebted to all contributors for pausing in their research long enough to make their contributions on schedule. Special credit is due to all our colleagues who contributed by supplying figures and data or made material in press available to the various authors. A number of colleagues aided me in the review process and I am most sincerely thankful to them. Last but not least, I thank Penny, Artemis, Lisa, and Nikki for their love, patience, and understanding; the time I spent on this and other simultaneous projects has detracted from my precious time with them.

John G. Scandalios

1

Genetic Factors in Developmental Regulation

K. PAIGEN

It gradually dawned on us that what we eventually came to identify as a "differentiated" cell of given type, shape, structure, colour, and behaviour, was really just the last scene of a long play of interactions, the earlier stages of which had been hidden from us. We came to realize that, in dynamic terms, the so-called "stages" of the development of a cell, an embryo, or a disease, represented simply cross-sections through a continuous stream of pro-

PHYSIOLOGICAL GENETICS
Copyright © 1979 by Academic Press, Inc.

cesses, recorded by whatever methods of assay and portrayal—whether in pictures or in graphs—we had at hand. The picture of an overt "character" thus emerged as merely the residual index of prior covert dynamics; "form," as the registered outcome of formative processes; . . . "fibres," as the result of the recruitment, polymerization, alignment, and bundling of masses of anisodiametric macromolecules; and so forth.

Paul Weiss (1973)

I. INTRODUCTION

To describe the role of the genome in guiding development is in effect to ask what changes in developmental processes attend individual mutations. Two quite distinct frames of reference have been used for this purpose. One is to express the outcome in terms of nuclear differentiation— the changing pattern of gene activities that varies both temporally, as development proceeds, and spatially, from one cell type to another. The result of nuclear differentiation is that each cell type comes to produce its own characteristic array of RNA and protein molecules. The other frame of reference has been morphogenesis, or pattern formation, including both the development of visibly specialized cells and the morphological structures containing them. The two frames of reference are related by the obvious fact that morphogenesis is brought about by the characteristic complement of macromolecules each cell produces. As Paul Weiss has so evocatively stated it, form is the registered outcome of formative processes.

A major experimental dilemma has been the difficulty of deciding whether an observed developmental mutation affects the process of nuclear differentiation itself or the subsequent steps of morphogenesis. This difficulty has been especially acute for the group of mutants in which a normal act of differentiation fails, such as in the stunted wings of the *Drosophila* mutant *vestigial* (Lindsley and Grell, 1967) or the death of the visual cells in the mouse mutant *retinal degeneration* (Noell, 1958). For the great majority of these mutants it has not been possible to determine whether they are impaired in the structure of a protein required for morphogenesis or whether they are impaired in the genetic mechanism that arranges for adequate amounts of that protein to be present in the right cell at the right time. The development of a contractile muscle will fail whether an amino acid substitution in one of the myosin chains destroys its functional capacity or whether myosin is produced in inadequate amounts, or in the wrong cell, or at the wrong stage of development.

The reasons for this dilemma are not hard to find. The products of many structural genes are required to bring about even one morphogenetic event; each of these structural genes has a genetic regulatory apparatus associated with it, and the various proteins coded by these structural genes, as well as the metabolic products of the reactions they catalyze, react with each other in processes whose identities are unknown, much less their concentration dependence, stoichiometries, and cofactor requirements. This has made it quite difficult to identify the primary gene product affected in a given morphological mutant. Similarly, if several genes are identified as affecting the same morphogenetic step, there is no way to test whether they even involve the same or different gene products. Among the thousands of morphological mutants known, the affected protein has been identified for only a few. These, such as *rudimentary* in *Drosophila* (Rawls and Fristrom, 1975), have turned out to be simple enzyme deficiencies and relatively uninformative regarding the fundamental mechanisms of either nuclear differentiation or morphogenesis. Thus, the importance of morphological mutants has primarily been in defining morphogenetic interactions.

Two other groups of mutants have, however, provided some information regarding the role of the genome in guiding development. One group includes mutants altered in the developmental programming of individual proteins. Mutants of this kind have been described for maize, *Drosophila*, and mice, and the concept has developed that a set of genetic units exists, called temporal genes, whose function is the coding and expression of these programs (Paigen, 1964, 1971, 1977; Paigen and Ganschow, 1965). Temporal gene mutants appear to be specific in affecting only single proteins, and those known do not have grossly observable morphological consequences. These mutants are discussed in Sections II and III.

The second group includes mutants with a substitution of one morphogenetic pattern for another. These are the homeotic mutants of *Drosophila*, such as *antennapedia*, where a leg appears in place of an antenna (for reviews, see Oberlander, 1972; Postlethwait and Schneiderman, 1973; Gehring and Nöthiger, 1973; Ouweneel, 1976). Such mutants are largely confined to insects, and their existence reflects the unique metamorphosis of these organisms in which some adult body parts develop from a small number of discrete larval structures, the imaginal discs. Each disc normally gives rise to a fixed complement of adult structures, and the homeotic mutants are explicable as a partial substitution of one complement for another. These mutants are discussed in Section IV.

Before turning to the mutants, it is appropriate to consider the extent to which the sequence of nuclear differentiation is the result of an intrinsic nuclear programming machinery and the extent to which it is a sequential response to metabolic changes in the cell, each change inducing the next metabolic transition. It will be some time before a definite experimental decision can be made as to which is the predominant mechanism, or whether they participate relatively equally. But even now there are reasons to suspect that intrinsic nuclear programming is at least an appreciable part of the whole. This is encouraging, for it is apparent that if nuclear differentiation is largely or entirely directed by a complex sequence of metabolic interactions, the experimental effort required to reach an appreciable level of understanding will be immense. This is in contrast to the possibility that a much more discrete set of elements and rules defines the operation of a nuclear programming system.

Although the nature of any intrinsic programming mechanism is virtually unknown, especially at the molecular level, there is considerable independent evidence that such mechanisms do exist. It appears that encoded DNA is a machinery that can determine a later sequence of highly specific nuclear events and that this machinery operates relatively independently of metabolic processes. The pioneering studies of McClintock (1951, 1965, 1967), now corroborated and extended by others (for reviews, see Fincham, 1973; Fincham and Sastry, 1974), demonstrate that in both plants and animals a class of genetic ''controlling elements'' exists that is capable of programming future genetic events in nuclei.

The expression of controlling element action is generally bipartite, requiring both a distant locus initiating the genetic action and the presence of a suitably sensitive allele of the structural gene that will respond. That is, the DNA in or near the structural gene must be of such a nature that it can respond to the action of the distant element. Two types of control are seen. One is the continuing ability of the distant elements to modulate the activity of sensitive alleles of the structural gene. In this sense the distant element acts as a genetic regulator that functions in all cells. The other, more dramatic, capability is that of programming the later occurrence of somatic mutations in sensitive alleles of the structural gene, but not in other genes. These mutations occur only during certain stages of development and in some tissues. That the somatic mutations represent true changes in DNA structure is deduced from the observation that when they occur in tissues that eventually give rise to germ cells the new phenotypes are transmitted through meiosis to the next generation of offspring.

What is immediately relevant is that the existence and properties of controlling elements demand the presence of genomic components capable of determining, for specific genes, the subsequent timing and tissue location of somatic mutations. Once this larger fact was established, it became easier to consider the smaller notion that a system, similar or identical in composition to the controlling elements described by McClintock, might program the changes in levels of gene function that characterize differentiation.

The examination of mutants with altered developmental programs for specific gene products provides a preliminary comparison of the relative roles that intrinsic genetic programming and sequential metabolic responses may play. To the extent that sequential metabolic changes direct the course of nuclear differentiation, we anticipate that mutations at many sites will affect the ultimate programming of even a single enzyme. In addition to the metabolic changes in the cell lineage itself, each dependent on earlier metabolic events, there will be the changes induced by metabolic or hormonal signals reaching the cell from outside. These will involve multiple genetic functions. Conversely, we would expect a mutation altering an earlier metabolic process to have multiple consequences, and thus have pleiotropic effects on the programming of many enzymes. In general, we would also anticipate that mutations affecting enzyme programming through metabolic perturbations would be recessive. Experience with many mammalian mutants, both in humans and experimental animals, shows that while mutant alleles generally show codominant expression at the level of the primary gene product, physiological consequences are usually recessive, since 50% changes in protein activity are easily compensated for (Childs and Young, 1963; Harris, 1964; Paigen and Ganschow, 1965). For example, heterozygotes of human enzyme deficiencies characteristically exhibit a normal phenotype despite half normal enzyme levels, even though the homozygous deficiency may produce severe symptoms or be lethal.

Thus, to the extent that sequential metabolic processes are an important mechanism in determining the process of differentiation, we would anticipate that the programming of one enzyme will be affected by mutation at many sites in the genome; that mutation at any one site is likely to affect the programming of multiple enzymes; and that mutant alleles will show recessive–dominant inheritance patterns rather than codominant–additive inheritance. Although the list is still relatively short, the enzyme developmental mutants analyzed so far do not have properties matching these predictions. As detailed in Section II a very high proportion of mutants are located in close proximity to the structural gene of the enzyme being programmed. There is a great deal of specificity, so that even enzymes located in the same cellular organelle and whose activity changes are coordinate in development are under independent genetic control, and all of the systems tested show additive inheritance. To the extent that it is possible to generalize from the examples now available, it appears that a great deal of intrinsic programming occurs.

The simplest type of programming that can be considered is that of individual structural genes. The temporal gene mutants provide examples of this. The question then arises as to whether a higher level of programming also exists, one that chooses between ectoderm and endoderm, between becoming an erythroid stem cell or a myeloid stem cell. The alternative is to consider that these larger steps represent the summation of many individual decisions. The origins of this question are really historical. Classic descriptions and definitions of the problems of embryology and development were framed in morphological terms. One heritage of that is our tendency to consider development as a series of discontinuous steps leading to the formation of new cell types. In superimposing the past on the present, we are implicitly assuming that morphogenesis involves a simultaneous change in the functional state of many genes, a kind of regulation en bloc. Such a discontinuous step requires the existence of master controllers of some kind, agents capable of initiating changes at many genes simultaneously. Serious theoretical consideration has been given to models and mechanisms that might provide this ability (Britten and Davidson, 1969; Gierer, 1973; Brawerman, 1976), and the concept remains an important part of our contemporary thinking.

There is, however, no experimental requirement that master switches exist. It now appears that morphogenesis is not necessarily synchronous with, or even an accurate reflection of, nuclear differentiation. The appearance of morphologically distinct cells is not always correlated with the magnitude or extent of the changes in protein complement that differentiated cells undergo. The limited utility of morphological criteria as an estimate of nuclear differentiation is well illustrated by the behavior of early mammalian embryos. During the preimplantation development of the mouse there are two major changes in the complement of proteins being synthesized (Van Blerkom and Manes, 1977). However, both of these precede the first morphological differentiation that occurs, the separation of the morula cells into an outer trophectoderm and an inner cell mass. The trophectoderm will eventually give rise to the tissues that interface the embryo and the maternal uterus, and the inner cell mass will eventually give rise to the embryo proper. The differentiation into inner cell mass and trophectoderm is not accompanied by any major change in the pattern of protein synthesis. The two differentiated cell types each synthesize virtually the same complement of proteins as the other and the same complement of proteins that was synthesized by the progenitor cells from which they arose. Nevertheless, the concept of master switches remains an attractive one, and, if they do exist, the outstanding candidates for membership are the homeotic genes of *Drosophila*.

II. TEMPORAL GENE MUTANTS

Mutants altered in the developmental expression of defined proteins have been described in several organisms including maize, *Drosophila*, and mice. They lack morphological consequences. The experimental strategy they offer is the possibility of distinguishing mutations affecting the production of specific proteins from mutations affecting the functional properties of these proteins. With this comes the ability to focus on the genetic apparatus involved in developmental programming of gene activity. For each structural gene a limited set of genetic sites, called temporal genes (Paigen, 1964, 1971, 1977; Paigen and Ganschow, 1965), has been found to participate in developmental programming. The basic pattern appears to include interaction between a proximate element located adjacent to the structural gene and one or several distant elements located elsewhere.

A. Proximate Loci

1. Glucuronidase

a. Background. The first temporal gene mutant described, and the one most extensively studied, controls the development of β-glucuronidase in various tissues of mice. Animals homozygous for this mutation undergo programmed changes in β-glucuronidase activity occurring at a different age in each tissue. Regulation is expressed as a control of enzyme synthesis; whether it involves a modulation of mRNA activity is unknown. The changes are enzyme specific in the sense that other enzyme activities, and especially other lysosomal enzymes, are not affected.

The variant phenotype was originally discovered in the C3H strain of mice, and present evidence suggests that this strain is altered in two regions of DNA that influence enzyme activity. One change is the production of enzyme molecules with altered structure and increased thermolability; the other change is a new age-dependent program of enzyme synthesis. Each change is determined by a single genetic site, and both sites map within the same small region of chromosome 5. The thermolability mutation, together with other structural variants exhibiting altered electrophoretic mobility, define the structural gene *Gus* for this enzyme. The failure to observe recombination between the structural locus and the genetic determinant for temporal regulation indicates that age-dependent synthesis is programmed by a chromosomal site *Gut* that is in close proximity to the structural gene. The limitations of fine structure mapping in the mouse do not allow a distinction as to whether the *Gut* site is adjacent to or integral with the structural gene. Immediately, the crucial point is that some DNA sequence is capable of programming age-dependent enzyme synthesis. The problems involved in deciding where this DNA sequence is located are presented in Section III.

Gut was historically important in providing the first evidence that a specific locus can program the level of a single enzyme during development; its close proximity to the structural gene and the fact that it acts by regulating enzyme synthesis imply that the mechanism is not trivial.

b. Biochemistry and Genetics. Considerable information is now available on the structure and regulation of β-glucuronidase in mice (for reviews, see Paigen *et al.*, 1975; Lusis and Paigen, 1977; Swank *et al.*, 1978). The enzyme is a tetramer of molecular weight about 280,000 and has been purified to homogeneity (Tomino *et al.*, 1975). The structural gene for the enzyme is located near the distal end of chromosome 5 and directs the synthesis of a polypeptide chain of molecular weight from 70,000 to 75,000. From this primary polypeptide chain the cell produces two physically distinguishable tetrameric forms of the enzyme, known as L and X (Ganschow and Bunker, 1970; Swank and Paigen, 1973). L is the classic lysosomal form of the enzyme that is familiar to most investigators. Its subunit molecular weight is 71,000. X is an alternate tetrameric form that is anchored onto the membranes of endoplasmic reticulum through complexing with a second protein, egasyn (Swank and Paigen, 1973; Tomino and Paigen, 1975). X differs from L in being 2000–3000 daltons heavier and in having a higher isoelectric point. The presence of egasyn requires the function of the *Eg* gene on chromosome 8 (Ganschow and Paigen, 1967; Tomino and Paigen, 1975; Lusis *et al.*, 1977). The complexes with egasyn serve to anchor the enzyme to membranes. A variety of evidence confirms that these multiple intracellular forms of glucuronidase are, in fact, derived from a single structural gene. Other than the differences related to their intracellular location, the various enzyme forms are identical physically, chemically, catalytically, and immunologically. Most convincingly, genetic experiments show that the properties of all intracellular forms of the enzyme are determined by a single genetic site (Paigen, 1961b; Lalley and Shows, 1974). Substitution of one structural allele for another at the chromosome 5 site simultaneously alters the properties of the enzyme at both intracellular locations. Three alleles of the glucuronidase structural gene have been reported: Gus^b, the standard form of the enzyme; Gus^h, present in C3H mice and determining the thermolabile enzyme mentioned above (Paigen, 1961b); and Gus^a, an electrophoretically fast form (Lalley and Shows, 1974).

The same structural gene is expressed throughout development; mutation at the *Gus* locus affects the properties of the enzyme at all developmental stages, from the preimplantation embryo (Wudl and Chapman, 1976) through the neonatal and adult stages (Szoka, quoted in Paigen, 1977).

A regulatory locus *Gur* is located in close proximity to the structural gene (Swank *et al.*, 1973, 1978). *Gur* controls the responsiveness of the enzyme in kidney to induction by androgenic steroids. Like *Gut, Gur* regulates the rate of enzyme synthesis. In heterozygotes that carry a different *Gur* allele attached to each copy of the structural gene, regulation is cis, that is, each chromosome is regulated according to the *Gur* allele it carries and acts independently of the other.

    c. **Developmental Regulation.** In addition to carrying the mutation affecting enzyme structure, C3H mice have an altered developmental program for β-glucuronidase (Paigen, 1961a). Relative to other inbred strains, such as C57BL/6, C3H mice undergo a relative decline in enzyme activity at an age that is characteristic of each tissue (Fig. 1). Thus, liver activity drops abruptly beginning at 12–15 days of age, whereas the decline in brain begins prenatally.

Three lines of evidence suggest that this is a unit phenotype, determined by a single genetic locus that is located in close proximity to the structural gene *Gus*. The first is provided by the results of conventional crosses. Because the decline in activity in C3H produces animals that are relatively enzyme deficient as adults, having only 5–10% the activity of most other inbred strains, progeny of test crosses can be scored for both the end result of the developmental phenotype and the structural form of the enzyme present (Paigen, 1961a, 1977). The results show complete cosegregation of the two phenotypes, with no recombination in several hundred tested chromosomes. Moreover, assays of segregating populations at various stages of development indicate that the enzyme levels measured in adults do reflect the end result of distinct developmental phenotypes.

The second line of evidence takes advantage of a special line of C57BL/6J mice, developed by Dr. E. Russell, that are congenic for a region near the distal end of chromosome 5. That is, this subline contains a C57BL/6 genome except for the region including the genes *light ear* and *Gus* on chromosome 5. This latter region has been derived from C3H mice. The developmental program for β-glucuronidase in these congenic mice is identical with that in C3H and quite unlike that seen in C57BL/6 mice that do not carry this substitution (Paigen, 1977). The congenic line demonstrates that this short chromosomal segment contains all of the genetic factors that distinguish the developmental phenotypes of C3H and C57BL/6 mice, as well as confirming the chromosomal location of the *Gut*

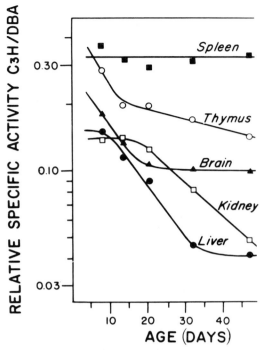

Fig. 1. Developmental programming by the *Gut* temporal gene is tissue specific. The relative activity of β-glucuronidase in C3H mice changes at different times for each organ. The data are plotted as the ratio of C3H specific activity to the specific activity of the corresponding organ in the reference strain DBA/LiHa at the same age. (Data from Paigen, 1961a.)

locus. V. Chapman (personal communication) has succeeded in reducing the size of this fragment further by recombination and demonstrated that *Gut* continues to remain closely attached to *Gus*.

The use of recombinant inbred lines derived from a cross of C3H × C57BL/6 has provided the third line of evidence. Recombinant inbred lines are sets of new homozygous inbred strains derived by inbreeding the F_2 generation of a cross between two different progenitor inbred strains. Various recombinant chromosomes and chromosome combinations present in the F_2 now become fixed in homozygous form in the new strains. Such a set of strains provides a powerful test for linkage of genetic determinants and whether a phenotype is influenced by more than one locus (Swank and Bailey, 1973). Using the set of lines derived from the progenitor strains C3H and C57BL/6, G. Breen and K. Paigen (unpublished) showed complete cosegregation of *Gus* and *Gut*.

Alleles of the *Gut* locus show additive inheritance. Both F_1 heterozygotes and heterozygotes arising in backcross populations have enzyme levels intermediate between those of their two parents* (Paigen, 1977). Whether this is because the *Gut* site, like *Gur*, acts cis, and the intermediate level reflects the average of two independently regulated chromosomes, or whether *Gut* acts trans, and regulation itself is intermediate is uncertain. Unambiguous tests of cis versus trans action have been difficult in this system. Herrup and Mullen (1977), using heterozygotes carrying Gus^b and Gus^h structural alleles and measuring thermolability of the mixture of enzyme heteropolymers produced, have concluded regulation is trans. However, A. J. Lusis (personal communication), using Gus^a/Gus^h heterozygotes and examining electrophoretic mobility of the enzyme produced, has not found an equal contribution from both chromosomes.

Gut control has recently been shown to act quite early in development, as well as at postnatal stages. Wudl and Chapman (1976) examined *Gut* expression in the first few days of embryogenesis using a sensitive fluorescence assay to determine enzyme levels in individual preimplantation embryos of different genotypes. In embryos of C57BL/6 there is a dramatic rise in β-glucuronidase beginning at about the 8-cell stage, with a more than 100-fold increase in activity over the next few days. This increase is both delayed and diminished when the Gut^h allele from C3H is introduced into the strain.

Mutation of the *Gut* locus specifically affects β-glucuronidase and not other lysosomal enzymes, despite the fact that the developmental profiles of lysosomal enzymes are coordinate. Meisler and Paigen (1972) have shown that β-glucuronidase and β-galactosidase activities change in parallel during the development of various organs, although the relative amounts of the two enzymes vary from one organ to another (Fig. 2). Studies of α-galactosidase indicate that it also shares the same developmental profile (Lusis and Paigen, 1975). Nevertheless, the Gut^h mutation in C3H mice affects only β-glucuronidase and does not alter the developmental profiles of the other two enzymes (Felton *et al.*, 1974; Lusis and Paigen, 1975). In a similar fashion, temporal gene mutations affecting de-

---

* Curiously, dominant expression of high activity was reported for the original cross defining the existence of a locus affecting β-glucuronidase activity in the C3H strain of mice (Law *et al.*, 1952). In this laboratory crosses of C3H with both DBA/LiHa, a Gus^a strain, and C57BL/6, a Gus^b strain, have shown additive inheritance. It is possible that this difference somehow reflects the exact genetic constitution of the strains used in the crosses. Alternatively, it may reflect some of the technical problems of quantitating glucuronidase that existed in the early days of work on this enzyme (see Paigen, 1961b, for a discussion of this last factor).

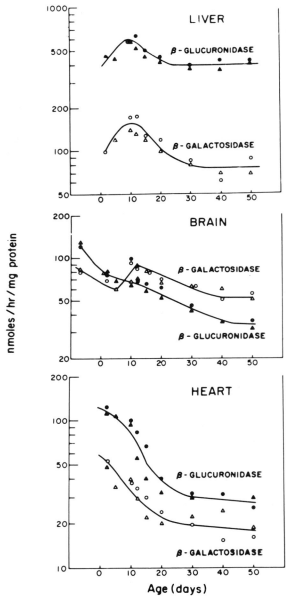

Fig. 2. β-Glucuronidase and β-galactosidase activities change coordinately during mouse development. Enzyme-specific activities are plotted for β-glucuronidase (closed symbols) and β-galactosidase (open symbols) in several organs of strains DBA/2J (circles) and C57BL/KsJ (triangles). (From Meisler and Paigen, 1972; copyright 1972 by the American Association for the Advancement of Science.)

velopmental programs for α-galactosidase and β-galactosidase are also independent (see Sections II,B and II,C). A curious and intriguing observation is that the developmental profile of β-hexosaminidase in a number of mouse strains resembles that of the Gut^h β-glucuronidase mutant (R. W. Elliott, personal communication), suggesting that only a limited set of developmental phenotypes may be possible among lysosomal enzymes.

The molecular basis for regulation by the Gut temporal gene has been examined in some detail by Ganschow (1975). The enzyme concentration of a tissue reflects a balance between the rate of enzyme synthesis and the rate of degradation by protein turnover. In the case of Gut gene control enzyme changes reflect changes in rates of enzyme synthesis. Several organs were examined before and after developmental changes, and in each case an abrupt change in the rate of enzyme synthesis was responsible for the mutant phenotype. The fact that Gut control is not a posttranslational or epigenetic process, together with the close proximity of Gut to the enzyme structural gene, suggests that Gut control may lie in the regulation of transcription. However, other mechanisms involving age-dependent changes in messenger structure are still possible, such as changes in messenger processing or translation efficiency.

   d. Conclusions. The major conclusions drawn from the experiments on β-glucuronidase are (1) the existence of a specific locus determining the developmental program of the enzyme, (2) the close proximity of the temporal site to the structural gene, (3) its additive level of control; (4) the tissue and enzyme specificity of its action, and (5) the fact that control is not posttranslational, but affects the rate of enzyme synthesis. Taken together these facts suggest the existence of a specific genetic regulatory process that functions to determine the glucuronidase developmental program, probably through tissue-specific and age-dependent changes in glucuronidase messenger activity.

2. Esterase

   a. Background. Various aspects of the biochemistry and genetics of maize esterase E_1 have been described by Schwartz and co-workers. This included the discovery of unusual alleles of the structural gene, called "prime" alleles, that cause precocious loss of enzyme in developing endosperm, but not other tissues. The genetic site responsible for the "prime" character proved to be very closely linked to the structural gene and to act in a cis fashion. The similarities to the glucuronidase system were marked, and esterase provided the first confirmation of a proximate temporal locus as well as the first demonstration of cis action. Prime alleles with the same timing characteristics were found in association with different structural alleles, and the same structural allele was found associated with various alleles of the "prime" site. Although not originally viewed in that light by Schwartz, the data suggest that the prime and structural sites are functionally distinct and probably separate, despite their close linkage.

b. Biochemistry and Genetics. The E_1 esterase of maize is a 95,000 dalton neutral esterase (Schwartz, 1964a). A total of seven electrophoretic forms of the enzyme have been described, each determined by an allele of the E_1 locus (Schwartz *et al.,* 1965). The electrophoretic forms fall into two families, one including the S, N, and F forms and the other the L, R, T, and W forms. Within each family the members appear to differ by unit changes; however, the mobilities of the S, N, and F forms do not coincide with the L, R, T, and W forms, and the nature of the interfamily difference is unknown. The enzyme is probably a dimer, since heterozygotes give a three-banded pattern consisting of both parental forms together with a more intense intermediate heteropolymer band (Schwartz, 1960).

The nature of the structural difference between members of the same family has been interpreted as a conformational change affecting exposed charges, rather than a simple amino acid substitution altering the net charge (Schwartz, 1967). The principal evidence for this conclusion is the observation that the SS homodimer behaves as though it is more negatively charged than FF at pH 8.5, and more positively charged than FF at pH 5.0, a fact that is difficult to reconcile with a simple amino substitution. In addition, all forms of the enzyme can be converted *in vitro* to a common electrophoretic form that is more negative by treatment with glyceraldehyde (Schwartz, 1965) or a combination of borohydride and an aldose sugar (Schwartz, 1964a,b). Treatment with iodoacetamide (but not iodoacetic acid) yields a common product that is more positive.

The E_1 esterase produced in different tissues appears to be derived from a single gene, since segregation at the E_1 locus determines the electrophoretic mobility of enzyme in both endosperm and seedlings (Schwartz, 1964b). However, the enzyme does not have the same molecular form in all tissues. The NN dimer, from endosperm, but not from seedlings, is sensitive to inactivation by 6 M urea (Schwartz, 1964c). This resistance of the seedling enzyme disappears after glyceraldehyde treatment (Endo and Schwartz, 1966). In contrast, the SS and FF dimers are resistant to urea, regardless of their tissue source and whether they are exposed to glyceraldehyde.

c. Developmental Regulation. E_1 esterase is normally present in endosperm in high amounts throughout kernel development. However, alleles of the E_1 locus have been found that cause enzyme deficiency after about 14–16 days of development (Schwartz, 1962). These developmentally unusual alleles have been termed "prime" alleles. The molecular form of the enzyme coded by the F' and S' alleles appears identical with that coded by the corresponding F and S alleles, as judged by electrophoretic mobility and sensitivity to inactivation. The genetic element responsible for precocious loss of enzyme is closely linked to the E_1 gene (0/2573 recombinants), and acts cis. Heterozygotes of F' with N or S exhibit a normal set of three enzyme bands in extracts of younger endosperm, but only the N or S forms of the enzyme are present in extracts of older endosperm. The developmental effect of the "prime" alleles is tissue specific. There is no loss of enzyme during the simultaneous development of the embryo itself or during subsequent development of the seedling after germination.

Eventually developmental mutants were found for alleles producing the F, N, S, and T forms of esterase (Schwartz, 1964d). These proved to be of two types with respect to endosperm development: those in which the precocious disappearance of enzyme was complete, and those in which it was only partial. All had normal levels during seedling development. In addition two very unusual N' alleles were found that showed a marked *elevation* of enzyme activity in aerial roots, a tissue unaffected by other "prime" mutants, in addition to a reduction in endosperm activity.

d. Conclusion. The "prime" mutants of E_1 esterase behave very much as though they are altered at a proximate temporal gene site closely linked to the enzyme structural gene. The mutant alleles act cis to cause a tissue-specific decline in enzyme activity during endosperm development. The proximate temporal gene site and the enzyme structural locus appear to be functionally distinct and physically separated, since the same mutant developmental phenotype is seen with different structural alleles, and the same structural allele may exhibit multiple developmental phenotypes. For example, at least three mutant developmental phenotypes, in addition to the normal one, were found in association with the N structural allele.

3. Aldehyde Oxidase

a. Background. Aldehyde oxidase provided the first evidence of a temporal gene system in *Drosophila*. A temporal genetic regulator closely linked to the structural gene determines the relative enzyme levels present at different stages in development. The control is quite similar to that seen with other proximate temporal elements. It shows additive inheritance and cis regulation.

b. Biochemistry and Genetics. Aldehyde oxidase has been partially purified from *Drosophila* (Dickinson, 1970). The inheritance pattern of electrophoretic variants suggests that the enzyme is dimeric. Electrophoretic variants were first identified in *Drosophila simulans* by Courtwright (1967) who used them to map the structural gene. Both electrophoretic and null mutants have also been identified in *Drosophila melanogaster* by Dickinson (1970) who has mapped the structural gene at $3 - 56.6 \pm 0.7$. Both null and electrophoretic mutants map at the same position; there is no complementation between mutants, and heterozygotes between null and enzyme positive mutants have 50% of normal enzyme levels. Neither of the null mutants possess antigenic cross reacting material.

c. Developmental Regulation. Enzyme levels are low during larval stages of *Drosophila* development, rise to intermediate levels after pupation, and rise again shortly before eclosion to give highest levels in newly emerged adult flies. Although most stocks have very similar levels of adult enzyme, there is considerable variation for levels of pupal enzyme. One group of stocks has approximately 45–55% of the adult level, and another group 85–95% of the adult level. These differences in relative activity of pupal enzyme to adult enzyme probably reflect tissue-specific regulatory factors. Although the tissue distribution of the pupal enzyme has not been reported, it is known that there are appreciable changes in distribution during development (Dickinson, 1971). The larval enzyme is predominantly located in hypodermis, and appreciable amounts are also present in malphighian tubules and in the gut. Enzyme distribution is highly localized. For example, the most anterior and posterior parts of the hindgut stain intensely for enzyme, whereas the central portion is devoid of enzyme except for two thin lines of enzyme-positive cells connecting the two ends. In the adult there is little or no enzyme in hypodermis; enzyme is present in malphighian tubules and gut, where its distribution is different from that of the larval enzyme, and in most male and female genital structures.

Crosses have been made between strains with high or low pupal to adult ratios, and also differing in the electrophoretic allele of the structural gene they carry (Dickinson, 1972, 1975). The results show that pupal enzyme levels are determined by a single Mendelian locus that is closely linked to the structural gene and shows additive inheritance. A low apparent recombination frequency (approximately 2%) was observed between this temporal gene and the enzyme structural gene. However, linkage is probably much tighter than is indicated by these results. Only one of the two possible recombinant classes was observed, and this was the class that could represent errors in scoring enzyme activity. Indeed, there was internal evidence that the ostensible recombinants probably represented errors in diagnosing the quantitative phenotypes. Tentatively, Dickinson has concluded that very tight linkage between the structural gene and its temporal regulatory site is probably the case.

The distribution of the temporal regulator among *Drosophila* stocks is not concordant with any particular structural allele. Each developmental class includes a variety of electrophoretic phenotypes, and strains with the same electrophoretic phenotype may belong to different developmental classes. Enzyme from all strains had the same rate of heat denaturation and substrate affinity.

Immunological methods were used to investigate whether the alterations in enzyme level reflected changes in the number or efficiency of enzyme molecules. Enzyme from adults of all strains showed the same antigenic equivalence, expressed in enzyme units, indicating that all strains produce enzyme molecules with the same catalytic efficiency in adults. The same equivalence was observed for pupal enzyme in two strains with high pupal enzyme levels and one strain with low pupal enzyme level, but one of the low strains produced pupal enzyme molecules with approximately double the catalytic efficiency of other strains and of the adult enzyme produced later in the same strain. The reasons for this last observation are unknown, but tentatively it may be concluded that temporal regulation here is exerted on the number of enzyme molecules present and does not represent a change in the quality of the enzyme produced.

Using animals heterozygous for an electrophoretic variant of the structural gene as well as for the temporal regulator controlling pupal enzyme levels a cis–trans test was conducted. Apparently the temporal regulator acts cis, since there was a predominance of pupal subunits representing the structural allele derived from the parent with high pupal activity.

d. Conclusions. The *Aldox* system strongly resembles the other structural genes described with adjacent temporal regulatory elements, reiterating the pattern of a stage- and tissue-specific regulation of enzyme activity through a control of the number of enzyme molecules present. Detailed biochemical studies have not progressed far enough in this system to confirm whether the level of control is on protein synthesis or is posttranslational, but the observations of close linkage and additive inheritance suggests that it is probably at the level of protein synthesis. No distant temporal regulators have yet been identified in this system.

B. Distant Loci

α-Galactosidase

a. Background. The first clear evidence for temporal genes separate from the structural genes they control was obtained in the mouse in studies of α-galactosidase and H-2 antigen. In the case of α-galactosidase, the temporal locus proved to be autosomal, whereas the structural gene is located on the X chromosome. The surprising discovery in this system was the finding that alleles of the temporal gene show additive inheritance, despite the fact that the locus is not a closely linked cis-acting regulator.

b. Biochemistry and Genetics. Murine α-galactosidase is remarkably similar to β-galactosidase in a number of respects (Lusis and Paigen, 1976). The enzyme reversibly associates into a tetramer at acid pH's, a dimer at neutral pH's, and a monomer in the presence of very low salt concentrations at neutral pH. The subunit molecular weight is approximately 70,000. It is differentially sialylated in various organs, with the highest levels of sialylation seen in liver and the lowest in kidney. An appreciable percentage of the kidney enzyme is bound to microsomal membranes, while the liver enzyme is entirely lysosomal. Like the α-galactosidases of other species (Kint, 1970; Rebourcet *et al.*, 1975) the structural gene of murine α-galactosidase is contained on the X chromosome (Kozak *et al.*, 1975; Lusis and West, 1976). The mouse structural gene is defined by a thermolability variant, and recent mapping data place the gene 9 centimorgans distal to *Mo* (Lusis and West, 1978).

c. Developmental Regulation. About 25 days of age, C57BL/6 mice begin increasing their level of liver α-galactosidase until they reach enzyme levels that are twice those seen in other strains (Lusis and Paigen, 1975). The increase is only seen in liver, and α-galactosidase levels in other organs of C57BL/6 mice do not change in comparison with other strains. This developmental increase in α-galactosidase is remarkably similar to the increase in β-galactosidase seen in the same strain (see Section II,C,1). Conventional crosses were consistent with the phenotype being determined by a single autosomal locus. Unexpectedly, heterozygotes showed additive inheritance. Despite the similarity in phenotypes of α-galactosidase and β-galactosidase, crosses of C57BL/6 with other strains showed no linkage between the temporal genes regulating the two galactosidases.

At the molecular level, temporal gene regulation was shown to involve the number, rather than the quality, of enzyme molecules present. Whether this control is exerted at the level of enzyme synthesis is unknown.

d. Conclusions. The importance of the α-galactosidase study was in providing some of the first evidence for the existence of a distant temporal gene and the observation of additive inheritance in heterozygotes carrying different regulatory alleles of this distant locus. This last point is of considerable significance, since additive inheritance is not easily reconciled with a regulatory system similar to those seen in microorganisms.

C. Systems with Both Proximate and Distant Loci

1. β-Galactosidase

a. Background. Studies of temporal gene programming of β-galactosidase in mice have provided the opportunity to demonstrate the existence in one system of both a proximate temporal gene, similar to the *Gut* locus of β-glucuronidase, and a distant temporal gene of the type seen with α-galactosidase. Although the developmental phenotypes seen with β-galactosidase are remarkably similar to β-glucuronidase and α-galactosidase, the three enzymes are under independent genetic control.

b. Biochemistry and Genetics. Acid β-galactosidase is present in nearly all mammalian tissues examined. It functions in the degradation of glycolipids ending in a terminal β-galactoside linkage, and a deficiency of the enzyme in humans is associated with ganglioside storage diseases (Suzuki and Suzuki, 1973; Wenger *et al.*, 1974).

The mouse enzyme has been purified to homogeneity (Tomino and Meisler, 1975). At neutral pH the enzyme is present as a dimer, containing identical subunits of molecular weight about 63,000. At acid pH the enzyme reversibly associates to form a tetramer. Like other mammalian acid hydrolases, β-galactosidase is a glycoprotein.

The enzyme undergoes tissue-specific processing, with respect to both subcellular localization and covalent modification. While the enzyme activity in mouse liver is confined almost entirely to the lysosomal fraction, about 30% of the activity in kidney is microsomal, the remainder being lysosomal. The enzyme from different tissues also differs in its content of sialic acid. The kidney enzyme is nearly devoid of sialic acid, whereas the liver enzyme contains an average of about four sialic acid residues per subunit (Lusis *et al.*, 1977).

Several observations indicate that, like β-glucuronidase, the enzyme in all mouse tissues and in both the microsomal and lysosomal fractions is derived from a single common structural gene. Most significantly, a single inherited mutation simultaneously alters the electrophoretic mobility of the enzyme from each of these sources (Lusis *et al.*, 1977). In addition, all forms of the enzyme cross-react immunologically with monospecific antibody prepared against purified liver enzyme, and they have similar kinetic and physical properties, including substrate affinity, dependence of activity upon pH, rates of temperature denaturation, and molecular weights (Lusis *et al.*, 1977; Tomino and Meisler, 1975).

Screens of inbred strains of mice resulted in the identification of several genetic elements involved in the expression of β-galactosidase activity. The structural gene *Bge* for the enzyme is located on chromosome 9 and is defined by an electrophoretic polymorphism. Two alleles of *Bge* have been found among a total of about 100 inbred strains surveyed, *Bge*[a] (fast mobility) and *Bge*[b] (slow mobility) (Breen *et al.*, 1977).

Located in the same region of chromosome 9 is a locus, *Bgs*, controlling systemic levels of β-galactosidase activity (Lundin and Seyedyazdani, 1973; Felton *et al.*, 1974). At all ages strains carrying the *Bgs*[h] allele have about twice the activity of strains carrying the *Bgs*[d] allele in all tissues that have been examined. *Bgs* shows additive inheritance, as *Bgs*[d/h] heterozygotes contain intermediate levels of enzyme. Although *Bgs* and *Bge* are tightly linked (0/163 recombinants), the two phenotypes apparently do not result from the same genetic alteration, since their distributions among inbred strains are not concordant (Breen *et al.*, 1977).

c. Developmental Regulation. Mice of strain C57BL/6 undergo a doubling of liver β-galactosidase activity beginning at 25 days of age, leaving them with an elevated enzyme activity in liver, but not other organs, that persists through life (Fig. 3) (Paigen *et al.*, 1976). Other strains fail to do this. Genetic analysis indicates that for some strains, such as DBA/2 and CBA, the difference lies in a genetic site proximate to the enzyme structural gene, but that for others, such as BALB/c and C3H, the difference lies elsewhere in the genome. The existence of the proximate temporal site *Bgt* has been confirmed by several independent lines of evidence. In genetic crosses between C57BL/6 and DBA/2, the elevated liver phenotype segregates as a single Mendelian factor closely linked to the *Bge* – *Bgs* region containing the structural gene (Paigen *et al.*, 1976). The *Bgt* site cosegregates with this same region in recombinant inbred lines derived from a DBA/2 × C57BL/6 cross (Meisler, 1976). Transfer of the same chromosome region from CBA mice into C57BL/6 mice by repeated backcrossing to form a C57BL/6 congenic line results in cotransfer of the CBA developmental phenotype (F. Berger, personal communication). *Bgt* is distinct from *Bgs* and *Bge* genetically as well as functionally, since alleles of all three genes do not appear to show the same distribution among inbred strains (Breen *et al.*, 1977).

In addition to the *Bgt* site located in the region of the enzyme structural gene, evidence from other crosses indicates that at least one, and possibly several other, temporal genes participate in developmental programming of β-galactosidase (Berger *et al.*, 1979). This is most clearly illustrated in a comparison of C57BL/6 and C57BL/10 congenic lines carrying substitutions of the *Bgt* – *Bge* – *Bgs* region from CBA and BALB/c. In the former case developmental control is transferred with the inserted CBA chromo-

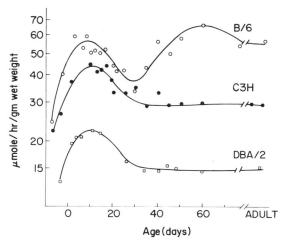

Fig. 3. Developmental and regulatory phenotypes determined by the temporal and systemic regulatory sites controlling the β-galactosidase structural gene. Enzyme-specific activities are plotted at different ages in livers of DBA/2, C3H, and C57BL/6 mice. The difference between DBA/2 and C3H illustrates the effect of the *Bgs* site, a systemic regulator causing a constant twofold difference in activity in all organs throughout development. The elevation in activity in C57BL/6 relative to the others that begins at 25 days of age illustrates the unusual developmental phenotype of this strain. The rise only occurs when correct alleles of both the proximate and distant temporal gene regulators of this enzyme are present. (From Paigen *et al.,* 1976, with permission of MIT Press; copyright © MIT.)

some segment, and in the latter case the BALB/c segment falls under a new control system and expresses a C57BL/6-like developmental phenotype (Table I). The suggestion that distant temporal genes determine the phenotype was confirmed in studies of BALB/c × C57BL/6 recombinant inbred lines, where new combinations of regulatory and structural phenotypes appeared (Fig. 4), and in conventional crosses of C57BL/6 with either BALB/c or C3H, where determinants for enzyme levels did not cosegregate with the *Bge–Bgs* segment containing the structural gene.

At the molecular level, immunological studies have shown that in all strains control of enzyme activity involves regulating the number of enzyme molecules present (Felton *et al.,* 1974; Paigen *et al.,* 1976). There are no differences between various mouse strains in enzyme activity per molecule, nor are there any differences within a strain before and after the developmental switch. Thus, the changes in activity determined by the systemic regulator *Bgs,* as well as the developmental changes programmed by the interaction of distant sites with the proximate temporal gene *Bgt,* reflect changes in enzyme concentration. More recently pulse-

TABLE I β-Galactosidases Activity in Congenic Mice Carrying Substitutions of the Structural Gene Complex[a]

Strain	β-Galactosidase structural gene allele	β-Galactosidase activity	
		Brain	Liver
I. Recipient C57BL/6	b/b	6.2	18.9
Donor CBA	b/b	2.8	6.0
Congenic C57BL/6.CBA.Trf[a] Bgs[d]	b/b	2.3	5.0
II. Recipient C57BL/10	b/b	6.1	21.5
Donor BALB/c	a/a	6.2	11.8
Congenic C57BL/10.C(47N).Bgs[h]	a/a	5.6	22.1

[a] In example I the Bge−Bgs−Bgt complex from strain CBA was transferred onto a C57BL/6 background to form the resulting congenic strain. In example II the same region was transferred from strain BALB/c onto strain C57BL/10 to form another congenic. Strains C57BL/6 and C57BL/10 exhibit the same genotype and phenotype. In example I the resulting congenic has donorlike regulation, indicating that the genetic basis for the regulatory difference is linked to the structural gene in example I but not in example II. (Data from F. Berger, G. Breen, and K. Paigen, unpublished.)

labeling studies have shown that these differences in enzyme concentration reflect differences in the rate of enzyme synthesis (Berger *et al.*, 1978).

d. Conclusion. It appears that mouse strains DBA/2 and CBA lack a developmental increase in β-galactosidase that is seen in C57BL/6 mice because they are mutant at the *Bgt* locus located in close proximity to the structural gene, whereas the BALB/c strain lacks the same developmental rise because it is mutant at one or more unlinked genetic sites. This developmental control is achieved at the level of enzyme synthesis, not post-translational processing or degradation. The nature of the interaction between the proximate and distant control sites that achieves this regulation is unknown. A likely supposition is that the proximate *Bgt* site acts as the receptor for a developmental signal that is provided or modulated by the distant sites.

The importance of the β-galactosidase experiments has been in (1) confirming the existence of proximate temporal genes in a case where the proximate element appears to be genetically as well as functionally distinct from the structural gene; (2) suggesting that both distant and proximate sites can exist in the same system with the possibility of studying interaction between them; and (3) demonstrating that developmental regulation of the three lysosomal enzymes studied (β-glucuronidase, α-galactosidase, and β-galactosidase) is genetically independent, despite the considerable similarity of phenotypes.

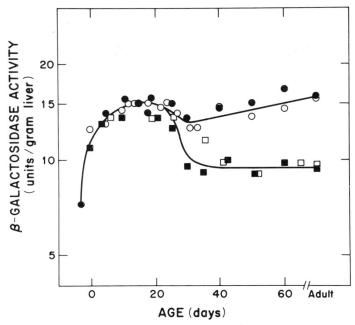

Fig. 4. Recombination between the genetic determinants for β-galactosidase structure and developmental phenotype in liver among recombinant inbred lines. β-Galactosidase specific activity of developing liver is plotted for the two progenitor lines BALB/c (closed squares) and C57BL/6 (closed circles), together with two of the recombinant inbred lines that each received the BALB/c form of the structural gene. One line (open squares) shows BALB/c type development, but the other (open circles) shows C57BL/6 type development.

2. Amylase

a. Background. *Drosophila* amylase has provided the most definitive evidence for the participation of distant temporal genes in the developmental regulation of an enzyme. During larval stages of development, amylase usually is present in both the anterior and posterior portions of the midgut. After metamorphosis, adult flies of all strains eventually come to possess enzyme in both regions. However, the time at which it appears in each region varies from one strain to another. In some, a full enzyme complement is already present at the time adult flies emerge. In other strains only some regions possess enzyme at emergence, and it does not appear elsewhere until considerably later. This spatial and temporal pattern of enzyme appearance is determined by at least two genetic sites that interact, one in close proximity to the structural gene for amylase, and another, *map*, some 2 centimorgans away. The proximate site appears to act cis, and the distant site to act trans.

b. Biochemistry and Genetics. *Drosophila* amylase is a neutral hydrolase that functions in the digestion of starch and glycogen by hydrolyzing internal α-1,4-glucosidic bonds. It has been characterized by Doane (1967b, 1969a) and was later purified by taking advantage of its specific binding to glycogen (Doane *et al.*, 1975). The purified enzyme is a monomer of 64,500 daltons.

The first genetic studies of *Drosophila* amylase examined crosses between strains of *D. melanogaster* varying in total activity (Abe, 1958; Kikkawa, 1960). The differences in enzyme activity proved to be determined by a single locus exhibiting additive inheritance and located on the right arm of chromosome 2. Subsequently, a series of electrophoretic variants were identified in *D. melanogaster* and mapped to the same position, suggesting that this is the location of the structural gene (Kikkawa, 1964; Doane, 1965, 1967a). The most recent evidence places the *Amylase* locus to the right of section 54B (Bahn, 1971). A total of six electrophoretic forms of the enzyme have been recognized, numbered according to their electrophoretic mobility. The *Amy* locus in *D. hydei* is located on chromosome 5 in the homologous chromosome segment (Doane, 1971; Doane *et al.*, 1975).

There is considerable evidence that the *Amylase* locus in *D. melanogaster* is actually a duplication with two structural genes in tandem. Electrophoretic variants apparently occur in both genes with the result that an individual fly may produce anywhere from one to four distinct amylases depending on its genetic constitution. The first observation suggesting the existence of a duplication was the presence of two electrophoretically distinct forms of amylase in single homozygous flies, with cosegregation of both electrophoretic phenotypes. Later Bahn (1968) confirmed this supposition directly by examining recombinants between two closely linked flanking markers, *curved* and *welt*. He showed that a small fraction of recombinants between these two markers were in fact recombinant for amylase alleles. Thus a fly that was Amy^1, Amy^1 on one chromosome and Amy^2, Amy^3 on the other chromosome could on rare occasion give rise to recombinant chromosomes that contained Amy^3 linked to Amy^1 or Amy^2 linked to Amy^1. In all, seven recombinants were obtained for an estimated map distance of 0.008 centimorgans between the tandem genes. A duplication may also be present in *D. hydei*. Homozygous flies express only a single electrophoretic form of the enzyme in somatic tissues (Doane, 1969b), but do express a second form of the enzyme in male sex glands (Doane *et al.*, 1975).

The original quantitative polymorphism is related to the electrophoretic polymorphism by the fact that various electrophoretic alleles are associated with different levels of enzyme activity, although the correlation is not absolute. The total range in activity between strains is approximately sevenfold, with the highest activities associated with the presence of the Amy^6 allele. However, comparison of enzyme levels in flies carrying various pairs of amylase alleles suggests that a simple one to one relationship between electrophoretic forms and enzyme levels does not occur. Instead, the two amylase structural genes appear to have somewhat different regulatory properties despite the fact that they can code for enzyme molecules that ostensibly have the same electrophoretic phenotype.

Enzyme levels are also regulated by the nature of the carbohydrate present in the diet of the flies (Abe, 1958; Doane, 1969b). The presence of high levels of starch increases enzyme activity, and this is suppressed by the addition of sucrose. In *D. hydei,* where the same enzyme induction occurs and where the chromosome material is favorable for cytological observation, it has been shown that increased enzyme activity is associated with chromosome puffing at a specific site in induced cells.

c. Developmental Regulation. Amylase is present in the midgut of both larval and adult flies, being located in the anterior and posterior regions of the midgut, but not in the central portion. All *Drosophila* strains have the same enzyme pattern in their larvae, but strains differ markedly in the adult pattern (Abraham and Doane, 1976, 1978; Doane, 1977). The spatial pattern of enzyme activity is easily demonstrated by dissecting out the entire midgut and laying it out on a film of starch gel. When left to air dry the cells break and liberate enzyme that hydrolyzes starch in the surrounding gel. Subsequent staining of the starch gel with iodine reveals the patterns of enzyme activity. Three general patterns of enzyme distribution in newly emerged flies are reported. In each case flies possess anterior region enzyme and the differences lie in the posterior region of the midgut. Flies with the A pattern possess enzyme in nearly all of the posterior region; flies with the C pattern lack enzyme in the entire posterior region, or are extremely reduced for it; and flies with the B pattern possess enzyme in only a small portion of the posterior region (Fig. 5). Flies that initially exhibit the B and C patterns eventually come to possess enzyme throughout the posterior region of the midgut, so that by 2–3 weeks of age the activity distribution in all flies looks alike (Fig. 6). The important distinction is the time during fly development when enzyme appears.

The genetic basis for the difference between the A and C patterns has been extensively investigated, and preliminary evidence is available regarding the genetic determination of the B pattern. The amylase activity present in both the anterior and posterior regions is derived from the same pair of structural genes. This was demonstrated by showing that in

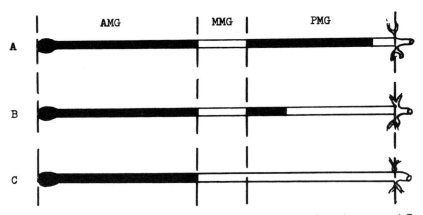

Fig. 5. Diagram of the amylase activity seen in midguts of newly emerged *Drosophila melanogaster* with the A, B, and C phenotypes. The C phenotype completely lacks enzyme in posterior midgut (PMG); the B phenotype lacks enzyme in the terminal region of the posterior midgut; and the A phenotype possesses enzyme throughout the posterior midgut. All flies show enzyme in the anterior region (AMG) and lack it in the central region of the midgut (MMG). (From Abraham and Doane, 1978.)

crosses the electrophoretic phenotypes of the amylases present in the anterior and posterior portions of the midgut are always the same. Thus the difference between the A and C patterns reflects a differential regulation of the same structural gene in two tissues, and not the expression of distinct structural genes. The delayed appearance of enzyme in the posterior portion of the midgut is determined by a single gene that undergoes recombination with the structural gene. The locus, *map*, is 2 centimorgans to the right of the amylase structural gene pair and much further away than the distance between the two structural genes. Combining estimates of the total number of chromomeres detected cytologically in *Drosophila* (Lefevre, 1974) and the total number of map units defined by genetic experiments (Lindsley and Grell, 1967) would place *map* approximately 35–40 chromomeres away from the amylase structural genes. The present conception is that *map^a* and *map^c* are two alleles of this locus determining the A and C patterns, respectively. The B pattern is thought to arise from a third allele at this locus, *map^b*. The *map* gene acts trans in its determination of posterior midgut activity. Heterozygous flies carrying *Amy^{1,6}map^a* on one chromosome and *Amy^{2,3}map^c* on the other chromosome produce all four forms of the enzyme in both anterior and posterior midgut regions. Thus, in heterozygotes posterior midgut enzyme is derived from both parental chromosomes, and a *map^a* allele is capable of activating the opposite chromosome to produce enzyme.

Visual comparison of staining intensities of midguts from homozygous and heterozygous flies suggests that heterozygotes have an intermediate

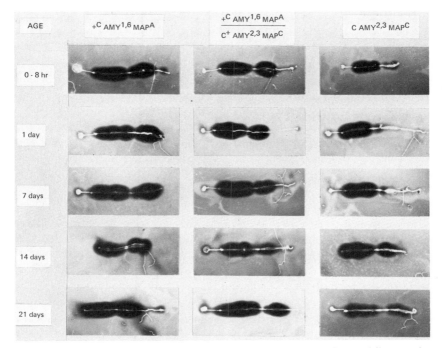

Fig. 6. Developmental appearance of amylase in midguts of *Drosophila* at various ages. Dissected tissues were air dried, laid on gels of starch, and regions of enzyme activity visualized as clear areas where starch has been hydrolyzed and no longer stains. It is apparent that flies with C phenotype eventually acquire enzyme in posterior midgut and come to resemble A-type flies. Heterozygotes possess enzyme early. (From Abraham and Doane, 1978.)

level of enzyme in posterior midgut, and that the *map* locus shows additive inheritance. That is, the amount of enzyme in the posterior midgut is proportional to the dosage of *map*^{a} alleles. In this respect the *map* locus is similar to the α-galactosidase temporal gene, which is located on an autosome, affects a structural gene on an X chromosome, and shows additive inheritance also. However, the conclusion that levels of posterior midgut amylase are proportional to *map*^{a} dosage is a tentative one, since quantitative measurements have not been made, and the situation is slightly complicated by the unusual response of the *Amy*^{2} allele as compared to other structural alleles.

In heterozygotes carrying *Amy*^{1,6} together with *map*^{a} on one chromosome and *Amy*^{2,3} together with *map*^{c} on the other chromosome, the four alleles contribute approximately equal amounts of enzyme in the anterior midgut. However, they do not do so in the posterior midgut, where the AMY-1, AMY-3, and AMY-6 enzyme forms are present in approximately equal amounts, but there is considerably less AMY-2 in young flies. The relative deficiency of AMY-2 in posterior, but not anterior, midgut suggests that genetic variation may exist in some kind of receptor element that is placed close to the structural gene and that acts to transduce the stimulus provided by *map*. The concept of a cis-acting element in close proximity to each copy of the structural gene agrees with the situation that has been described for other proximate temporal genes.

Preliminary evidence has also been obtained that the developmental programming of amylase in anterior midgut may involve an additional locus mapping in this region of chromosome 2R, a locus distinct from either the *map* or *Amy* genes. Very little is known about the developmental regulation of amylase in other tissues beyond the fact that the relative proportions of the amylase isozymes present vary from one tissue to another.

d. Conclusions. The *Drosophila* amylase system has been quite significant in providing the first definitive evidence for the existence and location of a separate temporal gene specifically affecting the development of an enzyme. Thus, DNA sequences distant from the structural gene in question also participate in programming enzyme appearance during development. These observations also have strong implications regarding possible mechanisms of action. Evidence from other systems, notably α-galactosidase, β-galactosidase, and H-2, also suggest the existence of separate temporal genes; however, in none of those systems has it yet been possible to definitively locate and map the distant element. This ability in the amylase system should make it possible to examine in a systematic way the interaction between distant and proximate temporal gene elements. The observations that have been made so far suggest that the *map* gene determines a nuclear environment in posterior midgut cells that allows or does not allow the expression of the *Amy* structural genes in these cells. The trans action of the *map* gene implies that the nature of that environmental stimulus is an interaction between some signal generated by the *map* gene and a proximate cis-acting receptor element located adjacent to each structural allele. The unexpected quality of these signals is their apparent additive influence. In both the amylase and α-galactosidase systems the final levels of enzyme found are approximately proportional to allele dosage, with heterozygotes showing intermediate enzyme levels.

3. H-2

a. Background. The developmental appearance of the H-2 antigen in mice is determined by the most complex system of temporal genes known (Boubelik *et al.*, 1975). The genetic model that was eventually developed to describe this system is a very provocative one, with many implications for other systems. It involves both a proximate cis-acting element, that itself has timing properties, and a pair of distant genes that may or may not take over the timing of H-2 appearance depending upon which alleles are present at all three sites.

b. Genetics and Biochemistry. The *H-2* region of the mouse, located near the centromeric end of chromosome 17, comprises the major histocompatibility complex of the mouse (for a general review of its genetics and properties, see Klein, 1975). Among other sites the region contains two genes, *H-2D* and *H-2K*, located at opposite ends of the region, that code for the structures of the major histocompatibility antigens of mouse cells. A large number of alternative genetic sequences of this region, or haplotypes, have been identified, analogous to the alleles of simpler genetic systems. In addition, extensive searches have been made for recombination events within the region. Breeding the recombinant chromosomes and testing for the separation of individual loci within the region has provided the beginnings of a fine structure map. The recombinants have also proved useful in testing the location of new genetic elements mapping within or near the *H-2* complex.

Considerable progress has been made recently in determining the structure of the H-2 antigens (see, for example, Grey *et al.*, 1973; Nakamuro *et al.*, 1973; Peterson *et al.*, 1974; Silver and Hood, 1974). Both H-2, and the homologous human protein HLA, are glycoproteins oriented in the cell membrane with an externally disposed amino-terminal chain containing the attached sugars. The antigenic determinants reside in the peptide sequence. There are interesting sequence homologies with sequences in immunoglobulin variable regions (Terhorst *et al.*, 1977). Associated with each H-2 chain of about 45,000 daltons is a 12,000 dalton β_2-microglobulin molecule.

c. Developmental Regulation. Virtually all cells acquire H-2 antigen quite early in embryonic development. The major exception is erythrocytes, which do not express H-2 until birth. All inbred mouse strains so far examined exhibit one of two phenotypes for the control of this timing (Boubelik *et al.*, 1975). In those with the "early" phenotype H-2 antigen is already expressed on their erythrocytes at birth, and in those with the "late" phenotype it does not appear until 3 days later. Initially, a variety of experiments using radiation chimeras (Boubelik and Lengerova, 1975) suggested that early versus late expression concerns the timing of appearance of H-2 antigen on erythrocytes, rather than nonspecific developmental differences. This conclusion was later confirmed in genetic experiments.

The genetic determinants for the early versus late phenotype do not appear to be identical with known elements of the *H-2* complex (Boubelik *et al.,* 1975). Strains carrying different *H-2* haplotypes often have the same developmental phenotype, and most importantly, strains carrying the same *H-2* haplotype may have different developmental phenotypes. This latter point is most obvious in the comparison of strains A (carrying *H-2ᵃ*) and B10.A (also carrying *H-2ᵃ*). B10.A was constructed by transferring the *H-2ᵃ* haplotype from strain A into strain B10 by a process of repeated backcrossing. The two *H-2ᵃ* haplotypes, therefore, have a common origin, and both strains synthesize antigen with the same specificity. Nevertheless, the *H-2ᵃ* antigen is expressed early in strain B10.A and late in strain A.

One of the genetic elements involved in determining developmental regulation of H-2 was identified in crosses of strain A, which has the phenotype H-2ᵃ-late, with strain A.By, which has the phenotype H-2ᵇ-early. In this cross the developmental phenotype was determined by a single Mendelian factor that proved to be closely linked to *H-2* (0/165 recombinants). The new locus, called Intrinsic (*Int*) acts cis in heterozygotes, which express their erythrocyte B antigen at birth and their A antigen only several days later. This cis expression confirms that H-2 developmental regulation is specific to this gene product and is not a general reflection of erythrocyte development. Despite the cis action of *Int* and the failure to observe recombination in a conventional cross, *Int* does not lie within the *H-2* complex as the complex is conventionally defined. Among the variety of recombinant chromosomes that have been isolated in the laboratory are several derived from cross-overs between *H-2D*, the rightmost marker in the *H-2* complex, and *tla,* a nearby marker several centimorgans distant. In these recombinants *Int* segregated with *tla* rather than with *H-2D*, showing that *Int* must be located an appreciable distance to the right of *H-2D*, despite its cis action. This location at the right, or *D* end, of *H-2* was confirmed by examination of mice carrying chromosomes that had undergone recombination within *H-2*. In these cases *Int* segregated with the *D* end of the region.

That *Int* is not the only gene participating in developmental regulation of H-2 became clear from a cross of strain A, which has the phenotype H-2ᵃ-late, with strain B10, which has the phenotype H-2ᵇ-early. Although these parents have the same phenotypes as in the previous cross, the results were very different. A fairly high frequency of recombination, some 16%, was observed between the genetic determinants for H-2 structure and timing. Examination of F_2 and backcross generations indicated a complex mode of inheritance. A genetic model capable of accounting for the data was eventually developed from an unusual type of cross. New inbred lines were established from the F_2 generation animals of the A × B10 cross by successive generations of brother–sister mating. Each line was selected for a reversed combination of parental phenotypes, either H-2[a]-early or H-2[b]-late. In the course of inbreeding the H-2 types were immediately established as homozygous, but the developmental phenotype continued to segregate for a number of generations, despite selection. By following the gradual disappearance of the phenotypes that were being selected against, Boubelik *et al.* (1975) were able to deduce a genetic structure compatible with the results. Their model received strong support when it proved an excellent predictor of the results of conventional crosses.

The model suggests the existence of two additional temporal genes *Tem* and *Rec* that are not on chromosome 17 where *H-2* is located. The two new genes reside on another chromosome, approximately 20 centimorgans apart. The interaction of the three genes *Int*, *Tem*, and *Rec* in determining the timing of H-2 expression is shown in Fig. 7. In the absence of suppression by the *Tem–Rec* system *Int* assumes control of H-2 timing and acts in a cis fashion. Two alleles of *Int* exist, one causing late and the other early expression of the adjacent *H-2* region. Suppression is determined by the nature of the *Rec* allele present. Two alleles of *Rec* occur, each matching one of the *Int* alleles. If a *Rec* allele matching the *Int* allele is present, then *Int* function is suppressed and *Tem* assumes control of timing. There are also two alleles of *Tem*, one causing early and the other late expression of H-2. In essence, the allele at *Int* determines the timing of H-2 unless a matching *Rec* allele is present, in which case *Tem* supersedes *Int* and assumes control of timing. One further constraint is required in order to account for the results. It is necessary to assume either that only the *Tem* allele cis to *Rec* can take over when a *Rec* allele suppresses *Int*, despite the fact that *Tem* and *Rec* are 20 centimorgans apart,

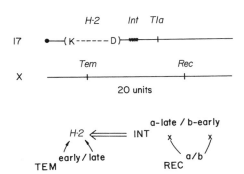

Fig. 7. Diagram of the genetic organization and interaction between the three genetic sites controlling developmental expression of H-2, *Int*, *Tem* and *Rec*, and the *H-2* structural region. (From Paigen, 1977.)

or more simply that *Tem* and *Rec* are located on the X chromosome, and show haploid expression because X chromosome inactivation leaves only one functional X chromosome in each cell.

d. Conclusions. The *H-2* system introduces several entirely new elements into our conception of temporal genes. One, is that despite the demonstrable cis action of *Int* on the regulation of *H-2*, *Int* and *H-2D* have clearly been separated by recombination. In this context we can estimate from the presently known size of the mouse genome (about 1250 centimorgans) that 1 centimorgan corresponds to a linear distance of approximately 1×10^6 to 3×10^6 base pairs. Cis control over distances of this magnitude requires unusual mechanisms of regulation.

The other novel factor introduced by the *H-2* system is the existence of *Rec*, a gene that suppresses the action of the proximate temporal gene *Int*. Whether *Rec* acts by inhibiting *Int*, and thereby allowing *Tem* control, or whether it provides an essential recognition component for *Tem* that allows the *Tem* product to function, is unknown. Whatever its detailed mechanism of action, the existence of *Rec* introduces the concept of suppression and the possibility that segregation at still a third site can determine which of two temporal genes will program a structural gene.

D. Miscellaneous Proteins

1. Aryl Sulfatase B

The predominant aryl sulfatase activity of mammals is the lysosomal aryl sulfatase B (Roy, 1958). Daniel (1976a) has described the structural gene *As-1* for this enzyme in mice. Two alleles are found determining alternate forms of the enzyme that are more or less thermolabile. Strain variation also occurs in kidney and liver levels of the enzyme (Daniel, 1976a,b). The kidney variation is present in males and apparently results from differences in androgen inducibility of the enzyme. The liver variation results from a developmental increase in enzyme after 35 days of age in some, but not all, strains. The two developmental profiles seen for aryl sulfatase B are very similar to the developmental profiles of other lysosomal enzymes.

Both the kidney variation and the developmental variation appear to be determined by single Mendelian loci. Both sites are linked to the enzyme structural gene that is defined by the thermolability polymorphism. Daniel (1976a) has reported recombination frequencies of approximately 10% between the structural gene and the kidney determining locus, and 25% between the structural gene and the liver determining locus *Asr-1*. These recombination values are uncertain, however, since tissue activity values did not segregate into nonoverlapping classes, and unambiguous assignment of animals to activity classes was not possible.

An interesting point is that the identity of the allele carried by a strain at *As-1*, the structural locus, is not correlated with the allele carried at *Asr-1*, the locus determining the developmental phenotype, supporting the view that these two are independent sites. Neither site is linked with *Gus*, the glucuronidase locus, or *Bgs*, the β-galactosidase locus, confirming the independent genetic determination of developmental programs for each lysosomal enzyme.

The work on aryl sulfatase B serves to confirm the conclusions reached in the other lysosomal ezyme systems (glucuronidase, α-galactosidase, and β-galactosidase) that genetic regulation of each lysosomal enzyme is independent of the others. The uncertainty in this system is the location of the temporal locus *Asr-1*, in particular, whether it is really separate from the enzyme structural gene.

2. Pyrimidine Degradation

Dagg *et al.* (1964) have reported a quantitative variation in rates of pyrimidine degradation between inbred strains of mice that results from a developmental difference. Pyrimidine-degrading capacity is virtually absent from liver at birth. Both SJL and C57BL/6 mice accumulate activity after that at the same rate. This stops after 6 days in C57BL/6 mice, but continues in SJL until a threefold higher level is reached. The difference was shown to be determined by a single Mendelian locus *Pd* showing additive inheritance. Individual phenotypes were determined by monitoring release of labeled CO_2 from animals injected with radioactive substrates.

Dagg *et al.* concluded that all three enzymes in the pyrimidine degradation pathway, dihydrouracil dehydrogenase, dihydropyrimidinase, and 3-ureidopropionase are affected by segregation at *Pd*. This was based on determinations of $^{14}CO_2$ released from liver homogenates exposed to substrates labeled with ^{14}C in the 2-position of uracil, dihydrouracil, and in the ureido carbon of ureidopropionate. Subsequently, Sanno *et al.* (1970) separated the three enzymes by column chromatography and assayed them individually. They report that only the last enzyme in the pathway, ureidopropionase, is reduced in activity in C57BL/6 mice and that the low activity strain also carries a structural mutation in this enzyme, decreasing its thermostability and affinity for substrate.

At the present time it is not certain whether the decreased activity seen in C57BL/6 mice results from a developmentally programmed premature reduction in enzyme synthesis or a more rapid rate of enzyme degradation. The latter could cause animals to have both lower enzyme levels and reach those levels sooner because of more rapid turnover.

3. α-Fetoprotein

Olsson *et al.* (1977) have described a developmental retention of α-feto-protein in the serum of BALB/c mice that is not seen in other strains. The difference is determined by a single recessive locus *Raf*. Measurements of clearance of labeled α-fetoprotein from blood suggested that the difference lies in the synthesis or release of α-fetoprotein into blood, rather than its clearance. The protein is produced in adult liver in very small amounts, presumably by the few dividing or embryonic type cells still present. Whether the *Raf* locus determines the proportion of these cells in the liver cell population, or the rate of synthesis per cell, is unknown.

4. *Drosophila* Enzymes

King (1969) has described a genetically determined developmental difference in xanthine dehydrogenase levels in *Drosophila*. In several strains of flies enzyme levels increase steadily in the days following emergence of adult flies at eclosion. However, unlike other strains the Sy strain only maintains this increase until the second day, at which time it begins a sharp decline in activity. Unfortunately, the genetic basis for this observation has not been determined. Xanthine dehydrogenase would appear to be a very promising system for analyzing temporal genes in light of the very large amount of work that has been expended on the detailed genetics of this enzyme in *Drosophila* (for example, see reviews by Glassman *et al.*, 1968; Dickinson and Sullivan, 1975; Chovnick *et al.*, 1977), including the important demonstration of a closely linked regulatory element by fine structure mapping (Chovnick *et al.*, 1976).

Genetic differences in developmental patterns of alkaline phosphatase (Johnson, 1966) and esterases, alcohol dehydrogenase, and octanol dehydrogenase (Gerdes, 1975) have also been reported in *Drosophila*.

III. PROPERTIES OF TEMPORAL GENES

The existence of a discrete, experimentally analyzable, genetic system for programming enzyme activities during differentiation is the first and most important conclusion that can be drawn from studies of the developmental genetics of single enzymes. Whatever the properties of this system may be, its existence is the preeminent fact that has been established so far. Its accessibility to study offers the ability to examine questions of development and differentiation from a new viewpoint. What then is known about the properties of this system of temporal genes, and what questions are raised by this information?

A. Stage and Tissue Specificity

Temporal gene expression is stage and cell type specific, which is less a description of properties than a definition of what constitutes a temporal gene. Loci such as *Lv,* affecting aminolevulinate dehydratase (Coleman, 1966; Doyle and Schimke, 1969), and *Bgs,* affecting β-galactosidase (Felton *et al.,* 1974), cause systemic changes in levels of enzyme equally at all ages and in all tissues. Such loci do not participate in developmental programming. In contrast, temporal loci such as *Gut, Bgt,* and *map* are tissue specific and act at fixed times.

The notions of spatial and temporal specificity are really reflections of the same basic property. For a mutation to affect adult enzyme levels in one tissue differently from another there must have been an earlier stage of development at which regulation in one cell lineage, but not the other, diverged from the norm. *Map^c* flies emerge lacking the posterior midgut amylase present in their *map^a* counterparts, although they possess enzyme in anterior midgut. A change in stage-specific regulation has occurred, the failure to synthesize enzyme in adult posterior midgut, that is also tissue specific, since anterior midgut has already initiated synthesis. We know this tissue-specific loss is really a timing difference since *map^c* flies eventually acquire the ability to produce amylase in the posterior midgut. If the delay in *map^c* flies exceeded their life span, and no posterior midgut amylase ever appeared, the notion of the timing component would be less obvious, but would still be required. In essence, the timing difference alters the relative properties of anterior and posterior midgut cells.

B. Proximate and Distant Loci

It seems likely that the aggregate temporal gene system for each enzyme is bipartite, including both proximate sites that are closely linked to the structural gene and distant sites that are clearly separate. This is the case for β-galactosidase, amylase, and H-2. The location of the distant sites ranges from 2 centimorgans away in the case of *map* and amylase, up to location on another chromosome altogether in the case of the temporal genes for α-galactosidase, H-2, and β-galactosidase. For glucuronidase in mice, aldehyde oxidase in *Drosophila,* and esterase in maize only proximate sites have been identified, and for α-galactosidase in mice only a distant site. There is no reason to believe that the failure to observe both proximate and distant sites in any one system reflects anything other than limited sampling. All of the known sites were identified in relatively small surveys that sampled some of the natural polymorphisms fixed during the establishment of inbred lines of various organisms. In no case has a selective system been devised that would permit surveying the total system of temporal gene elements regulating a given enzyme.

If the known collection of temporal genes is any reasonable reflection of the total polymorphism that exists, then the fact that proximate sites comprise a large fraction of all temporal genes so far identified suggests that each system is relatively simple in structure.

The high frequency of proximate sites was historically important in supporting the nontriviality of early observations on temporal genes. However, the existence of a high frequency of proximate sites brought with it an important question of mechanism. While it would seem that evolution would favor the existence of separate DNA sequences determining enzyme regulation and enzyme structure, permitting independent evolutionary selection of these two functions, it is also possible that the base changes affecting developmental profiles are actually changes in the coding sequences of the protein under control. This idea carries with it the implication that the amino acid sequence of an enzyme can somehow participate in determining its own developmental program of synthesis.

Initially, several factors made it difficult to resolve the question of where the proximate sites are located in relation to the structural gene. One has been the limitations of fine structure genetic mapping, especially in the mouse where many of the known systems are. Even very, very low recombination distances, well below 1 centimorgan, correspond to enormous molecular distances. Thus, any proximate temporal gene element close enough to its structural gene to exert cis control will be very difficult to separate by conventional recombination mapping procedures. A second factor is that many closely linked genes show linkage disequilibrium. That is, various alleles of closely linked genes are not associated at random; instead, certain groupings of alleles occur far more frequently in nature than others. One consequence of this is that when inbred strains of an organism are established, alleles of closely linked genes come to show a concordant distribution between strains, even when it is clear that the two genes are functionally distinct. For example, the genes for the β and y globin chains are distinct but closely linked structural genes, and the y^1 form of the hemoglobin y chain is always found in association with the Hbb^s allele of the β chain gene (Stern *et al.*, 1976). In a similar manner only certain pairings of structural alleles occur for the salivary and pancreatic amylases, whose structural genes are closely linked (Kaplan *et al.*, 1973; Nielsen and Sick, 1964). The same may apply to closely linked structural and regulatory genes. Thus, in the case of glucuronidase not only does the Gut^h allele of the temporal locus always occur in combination with Gus^h, but the Gur^a allele of the inducibility regulator gene almost always occurs in combination with Gus^a.

Several facts, however, suggest that proximate sites have a distinct temporal regulatory function and that they are not part of the coding sequence of the enzyme being regulated. In at least one case a proximate temporal gene has been separated by recombination from its closely linked structural gene. In the *H-2* system, a recombinant has been obtained separating the *Int* gene and *H-2D*, the locus it programs. This separation is all the more important since *Int* has been shown to act cis. There are, moreover, several systems in which linkage disequilibrium is not present. In these cases a given structural allele may be found attached to various alleles of its proximate temporal gene, and what appears to be the same allele of the temporal gene may be found associated with different structural alleles. This is true of the esterase E_1, β-galactosidase, and H-2 systems.

An important related observation has been made for *Drosophila* xanthine dehydrogenase. In that case a closely linked regulatory site has definitely been located outside the region coding for the amino acid sequence of the enzyme (Chovnick *et al.*, 1976). Although it is not known whether the site is a temporal, systemic, or some other form of regulator, the important principle has been established that regulatory sites are present in the DNA adjacent to the structural gene.

The second reason for suspecting that independent DNA sequences are used for developmental programming and structure determination comes from the existence of distant temporal genes. Their existence requires that programmatic information must exist outside the enzyme structural gene. Once we know that this occurs, we are relieved of the necessity, or even the advantage, of generating an additional and necessarily partial explanation for developmental programming by the proximate sites. It is easier to assume a unitary explanation. Moreover, the existence of distant sites carries with it the implication that they somehow transmit molecular signals to the chromosomal region containing the structural gene whose expression is influenced. Specific receptors for these signals must exist to ensure that they only act on the correct locus, and DNA sequences must exist that define their site of action. The proximate temporal gene elements are obvious candidates for these functions.

C. Enzyme Specificity

All of the temporal gene mutants described so far are enzyme specific, affecting only a single species of enzyme molecules. This is true both for those that map in close proximity to the enzyme structural gene and those that map at a distance. Obviously, in no case have all other enzymes been tested for possible effects. We do know, however, that the mutations do not produce gross tissue changes, nor, in the case of the temporal genes affecting lysosomal enzymes, do they affect other enzymes in the same intracellular organelle. For the proximate temporal genes this is hardly surprising, in view of their close association with a single structural gene and the fact that several are known to act cis. It is more significant for the distant regulators.

In the case of lysosomal enzymes we also know that there is no experimental bar to detecting polymorphism affecting more than one lysosomal enzyme. Swank and collaborators (1978) have described a system of genes altering tissue levels of lysosomal enzymes as a group; however, the mechanism is an effect on lysosome processing and excretion. This makes the failure to observe class controllers in the case of the mouse strain C57BL/6 all the more notable. This strain exhibits a liver-specific elevation of several lysosomal enzymes, but genetic analysis has shown that in each case it results from an independent set of temporal genes. The temporal gene regulators for one enzyme do not affect the others. Whether the coincident regulation of several enzymes in one strain results from coselection of lysosomal enzyme levels that are physiologically compatible with each other or whether it reflects the operation of some as yet unrecognized physiological influence remains unknown.

D. Molecular Level of Control

Much effort has gone into establishing the level at which temporal gene control is exerted along the chain of events leading from primary gene transcript to realized enzyme activity in the cytoplasm. Assuming that control is at a similar level in different systems, the data have served to limit progressively the possible area of regulation to somewhere in the control of the primary transcription–translation steps.

In several systems direct evidence has been obtained, using immunological titration, that genetically determined differences in enzyme activity reflect changes in the numbers of enzyme molecules present and not structural changes altering relative catalytic efficiencies. Data of this kind have been reported for β-glucuronidase, α-galactosidase, β-galactosidase, and aldehyde oxidase.

Changes in numbers of enzyme molecules can result from changes either in the rate of enzyme synthesis or the rate of enzyme degradation. A direct test of these possibilities was first achieved in the developmental control of β-glucuronidase by *Gut,* when Ganschow (1975) showed by pulse-labeling methods that regulation involves control of enzyme synthesis. Similar results have now been obtained in the temporal gene regulation of β-galactosidase (Berger *et al.,* 1978). The facts of the case also implicate regulation of synthesis for both mouse H-2 and *Drosophila* amylase, where temporal control determines when during development a protein appears in a specific cell. Finally, Dickinson (1975) has concluded that temporal regulation of aldehyde oxidase is not only cis but is probably a control of enzyme synthesis, from indirect arguments relating to the proportions of enzyme subunits present in flies heterozygous for both a structural allele and a linked temporal site. In no case has control been shown not to involve the synthetic process.

That other temporal gene loci also influence the events leading to enzyme synthesis is the most straightforward interpretation, but this must obviously be tested experimentally before it can be accepted. The immediate question in the case of *Gut,* and any other temporal genes that are known to control protein synthesis, is whether the regulation reflects changes in mRNA activity. If changes in mRNA activity do occur, this could be a consequence of a change in the number of mRNA molecules present, resulting from some alteration in the genetic control of mRNA production and degradation. Alternatively, it could result from a developmental change in the structure of the mRNA, bringing with it an altered efficiency of mRNA translation. A decision between these last two possibilities would have many implications regarding the basic molecular details of temporal gene regulation.

E. Additive Inheritance

The observation of additive inheritance for alleles of proximate temporal genes is to be expected if they act cis. The three closely linked regulators that have been tested, those for esterase E_1, aldehyde oxidase, and H-2, act in this manner, and it is likely from the difference in responses to *map* control by alleles of the amylase structural gene that a similar cis-acting element exists there also.

The observation of additive inheritance for a distant temporal gene in the case of α-galactosidase and probably *map* is a surprise and has significant implications. Virtually all genetic variants known in higher organisms that affect enzyme levels by acting posttranslationally show dominant–recessive inheritance. The mutations affecting the turnover system of catalase (Rechcigl and Heston, 1967; Ganschow and Schimke, 1969), complexing of β-glucuronidase with egasyn (Ganschow and Paigen, 1967; Karl and Chapman, 1974), and sialylation of several enzymes (Lalley and Shows, 1977; Dizik and Elliott, 1977, 1978) all have an F_1 phenotype that is identical to one of the homozygous parents. Thus, the observation of additive inheritance for alleles of the α-galactosidase temporal gene suggests that it does not act posttranslationally.

The further implication of additive inheritance is that if a gene does act in the regulation of enzyme synthesis, the molecular nature of its function in the control mechanism is unlike any of the functions seen in microbial systems, where mutations of unlinked regulatory genes are invariably recessive or dominant. Microbial trans-acting regulatory proteins, either repressors acting negatively or activators acting positively, react with an operatorlike DNA sequence to regulate an adjacent structural gene or group of genes. When different alleles of the genes producing the repressor or activator proteins are combined in the same cell, one allele is almost always dominant over the other. In each case, the reasons for this can be understood as coming either from the kinetic parameters that describe the interaction between the regulatory protein and its operator site of action or from the nature of subunit interaction between the monomers making up the regulatory protein. These structural and kinetic constraints are required to maintain a functional regulatory system. Thus, the observation of additive inheritance for regulatory elements in eukaryotes implies a very different molecular basis for control.

F. The Nature of Temporal Regulation

The existence of distant temporal genes regulating the output of structural genes located further along the same chromosome, or on another chromosome altogether, implies that some form of molecular signaling occurs. Very little is known about the nature of this signaling, and for the moment we cannot do much more than speculate. However, it is possible to frame some of the questions that must eventually be answered.

Factually, the evidence is that temporal gene control operates somewhere in the transcription–translation pathway leading to protein synthesis. Additionally, alleles of a distant regulator show additive inheritance in the case of α-galactosidase and probably also for *map* regulation of amylase. As was pointed out, this implies that the molecular signaling seen in the developmental regulatory system of higher eukaryotes is unlike that seen in microbial regulatory systems.

The most obvious question that can be asked about temporal gene signaling concerns the chemical nature of the signal itself. This must be a macromolecule to account for the specificity these signals exhibit, but there is presently no evidence to decide whether it is a protein, RNA, or even DNA. Considering that the signaling is intranuclear, that protein synthesis is cytoplasmic, and that a variety of large RNA molecules of uncertain function exist in the nucleus, it is tempting to consider that temporal gene signals are included in this population of large RNA molecules.

A second and more general question concerns the nature of the regulatory event itself. Two basically different kinds of control are possible. In one, regulation is dynamic, and depends upon the continued presence and concentration of the regulatory signal. In the other, regulation is structural, and the regulatory signal induces a change in the structure of DNA or chromatin that persists in the absence of the signal, perhaps until the next mitosis, or perhaps even longer. It is of course possible that some systems use one mechanism and some the other, but the unity of biology suggests that this may not be so. The importance of this question is large, and its answer is very fundamental to understanding the genetic control of differentiation.

Recently, attention has been drawn to the role of stochastic events in differentiation. Gusella and co-workers (1976) in an elegant series of experiments have introduced the idea that the commitment to differentiation by Friend erythroleukemia cells is probabilistic in nature. At each cell division in the parent culture a decision is made as to whether the two daughter cells will remain functional stem cells and capable of indefinite growth, or whether the two daughter cells will become committed to differentiation. If the two daughters do become committed, they will then initiate globin synthesis and undergo four subsequent divisions to produce 16 terminally differentiated erythrocytes each. The probability of commitment is normally low, and the role of chemical inducers, such as dimethyl sulfoxide (DMSO), is to increase this probability. The final probability is a function of DMSO concentration. If the probability rises above 0.5, then the entire culture will eventually become committed as the stem cells fail to replace their numbers. Similar arguments with regard to aging have been put forward by Holliday et al. (1977). They suggest that primary cell cultures do not have a fixed life span, but rather that at each cellular division there is a probability that the two daughter cells will become committed to a finite life span, undergo a fixed number of divisions, and then cease any further growth.

The significance of these observations lies in introducing the concept of stochastic decision making at each cell division. Whether elements of stochastic mechanisms are present in temporal gene control systems is unknown. Relevant to this is the fact that all of our present knowledge is based on the gross properties of tissues, and virtually nothing is known about the range of differentiated properties of individual cells in tissues. In this context, there may be an important lesson to be drawn from the historical precedent of microbial genetics. There, one of the most important conceptual steps in understanding molecular genetics was the introduction of a distinction between the phenotype of a mass culture and the separate phenotypes of the individual cells in it.

An interesting aspect of a possible relationship between stochastic decisions and temporal gene programming is that in our present ignorance either is a possible causal factor for the other. Stochastic events could participate in the basic mechanisms that provide the temporal and spatial specificity of developmental programming. Conversely, although there is only a limited probability that the daughters of an erythroid stem cell will differentiate into erythrocytes, the probability is zero that the daughters of a dividing intestinal crypt cell or skin fibroblast will do so. The erythroid stem cell has differentiated the capacity to undergo a probabilistic division with two alternate specialized consequences, and it has differentiated a machinery that allows this probability to respond to certain external influences. The developmental programming of these capacities may not be different in principle from the developmental programming of enzyme synthesis.

G. Questions of Generality

The question has sometimes been raised as to whether genetically determined developmental modulations of enzyme activity are somehow specialized and do not reflect basic mechanisms involved in cellular differentiation. Indeed, the question can be asked in several different senses: whether, because the magnitude of many of these changes is quite moderate, they somehow do not reflect the generality of control mechanisms; whether, because many of these changes involve "housekeeping" rather than "specialized" enzyme functions, they somehow do not reflect important events of differentiation; and whether, because the experimental studies have focused for the most part on the later stages of development, they somehow do not reflect the mechanisms operative in the early steps of cellular differentiation.

The issue of magnitude rests on the implicit assumption that one set of regulatory mechanisms controls large changes in activity and a very different set controls small changes. This seems most unlikely. Moreover, the known mutations affecting the developmental control of enzyme activity produce changes that vary from the barely detectable to the extreme case of presence versus absence. Because large changes are often lethal, or at least deleterious, most studies have relied on those smaller changes that are compatible with viability and a reasonable state of health of the organism. Occasionally larger changes, such as the mouse mutation affecting glucuronidase development or the *Drosophila* mutation affecting amylase, are compatible with viability and survival. When these have been studied, there is no suggestion that the mechanisms involved are any different, or that these mutations have occurred in qualitatively different genetic systems. At the present time, we have no reason to suspect that qualitatively different genetic systems are responsible for larger and smaller effects in the regulation of enzyme levels during development.

The issue of "housekeeping" versus "specialized" enzymes rests on the implicit assumption that two genetic systems of developmental regulation have arisen, one for housekeeping enzymes and the other for specialized enzymes. Several biological and evolutionary constraints make this most unlikely. The same enzyme may play a housekeeping function in one cell type and be a specialized enzyme in another. For example, the lysosomal enzymes are ubiquitously present in virtually all cell types. However, they have a specialized function in macrophages that act phagocytically to ingest and digest foreign materials. Similarly the P_{450} system of microsomal mixed-function oxygenases that oxygenate a variety of endogenous substrates are present at a low level in most cell types. However, they are especially abundant in liver, where they comprise upward of 25% of the total endoplasmic reticulum proteins and function to deal with the xenobiotics that enter the circulation via the hepatic portal system. It is difficult to accept the idea that for some enzymes two dissimilar regulatory systems are present side by side and that which system is used in a cell depends upon whether the enzyme's function can be described as housekeeping or specialized.

There are also evolutionary difficulties with a notion of two regulatory systems. For example, the rat, but not other rodents, makes enormous amounts of β-glucuronidase in its preputial glands. Nearly 5% of the total protein of the gland is this one enzyme, and much of the enzyme produced is released by secretion. This is surely extreme specialization, for there are few cells that spend a higher percentage of their total protein synthetic capacity on one protein. Yet, we would hesitate to suggest that in one step the rat has evolved an entirely new system for genetic regulation of glucuronidase in preputial glands. Instead, we expect that selection has chosen an exaggeration of preexisting regulatory mechanisms.

The extreme examples of specialized versus housekeeping functions are provided by proteins, such as hemoglobin and silk fibroin, that are only produced in a single class of cell, and then in massive amounts. It is worth pointing out that even this extreme synthesis is not achieved through any unusual transcriptional mechanism. Rather, it results from accumulation of stable messenger RNA over a period of time. Gland cells do not transcribe silk fibroin genes any faster than ribosomal RNA genes, for example (Suzuki et al., 1972). The one really distinctive developmental control that is known is the amplification of rDNA that occurs in amphibian oocytes in preparation for the massive ribosomal RNA synthesis that follows, and gene amplification does not occur in the ordinary course of differentiation.

The question of whether "late" differentiation steps are somehow different from those occurring during early development goes to the root of
our perception of what differentiation is. If by differentiation we mean the
appearance of persistent changes in the functional or morphological properties of cells, then there is no dichotomy between late and early events.
However, many biologists have an instinctive feeling that the more dramatic earlier steps (the first separation into recognizably distinct cell
types, the laying down of the primary germ layers, and the establishment
of organ anlage) are what differentiation is really about. Nevertheless,
considered reflection again suggests it is unlikely that life has evolved two
sets of genetic regulatory mechanisms, one for early modulations of nuclear differentiation and enzyme activity and another for later on. In one
case, it has been possible to put the question of early and late control to
experimental test and determine whether the same regulatory system
functions both early and late in development. The *Gut* locus controlling
developmental changes of β-glucuronidase was originally discovered and
defined by its action at terminal stages of hepatic development (Paigen,
1961a); recently Wudl and Chapman (1976) have shown that the same
locus acts as early as the 8-cell stage of mouse embryogenesis.

The quote from Paul Weiss which begins this chapter reminds us that
morphogenesis does not occur in the absence of biochemical changes,
that it is a reflection of underlying changes that began earlier in cells
that looked alike. The crucial problem is what initiates and regulates
these changes. In the unlikely event that two distinct genetic regulatory
mechanisms exist, one for early and one for late stages of development,
their equivalent importance in deciding the final phenotype of the organism, and the paucity of our knowledge about either, makes it a matter of
importance to investigate whichever is amenable to study.

H. Controlling Elements and Temporal Genes

There are important parallels between the class of controlling elements
first discovered by McClintock and the system of temporal genes. Both
involve changes at individual loci at fixed stages of development in specific tissues, and both are bipartite, apparently requiring interaction
between proximate and distant DNA sequences, presumably through molecular signaling. The very large question concerning the relationship
between the two is whether the changes induced are at all equivalent in
molecular terms. The developmentally timed events initiated by controlling elements are true mutations, transmissable through meiosis; the

timed events initiated by temporal gene systems are changes in levels of gene function, with no evidence that they involve DNA sequence changes. An answer as to whether the two can be equated will have to wait for clarification of the molecular nature of temporal gene control.

However, whether controlling elements and temporal gene systems are or are not equivalent, in whole or in part, the importance of controlling elements for our thinking about the programming of nuclear differentiation and the molecular activities of the nucleus has been large. At the minimum, the properties of controlling elements establish the existence of an intrinsic genetic apparatus, of extraordinary specificity, that appears to operate entirely independently of metabolic signals to produce nuclear changes at predetermined times in certain geographical locations. Even without knowledge of molecular mechanism, this precedent provides the confidence that intrinsic nuclear systems are capable of temporally and spatially specific programming of subsequent events.

I. The Coding Problem

The existence of programmatic information encoded in DNA raises some questions about how this is achieved. The first instance involves the chemical identity of temporal regulatory signals. If these are protein, then we can assume that the triplet code we know is the coding mechanism, in the same sense that it underlies microbial regulatory mechanisms. However, if the temporal regulatory signals are polynucleotide in nature, we should then ask whether there is any relationship between their sequence and their function, beyond the arbitrary necessity of determining specificity.

In a more general sense, however, whatever the nature of the signaling molecules may be, there exists the larger question of how to encode information instructing these genes when and where to function. The introduction of temporal and spatial factors as parameters of genetic regulation suggests that a developing organism possesses something that is the functional equivalent of a clock in its nuclei, together with some mechanism for telling time, as well as some orderly system for informing a given copy of a gene which cell it resides in.

As already mentioned, the requirement that genetic timing devices exist long precedes experiments on temporal gene regulation of specific enzymes. The concept is essential to explaining the controlling elements described by McClintock. How this information is encoded in DNA and how it is retrieved and expressed are profound questions.

IV. HOMEOTIC MUTANTS

A. Background

Among the large array of morphogenetic mutants that have been described among many organisms, the group of homeotic mutants of Diptera have received special attention. They are almost unique in representing a change from one differentiative capacity to another, rather than the loss of a normal function. For example, the homeotic mutant *antennapedia* (*Antp*) of *Drosophila*, causes the appearance of a leg in place of an antenna on the head (Fig. 8). (For a catalogue of homeotic mutants, see Gehring and Nöthiger, 1973; Postlethwait and Schneiderman, 1973.) A

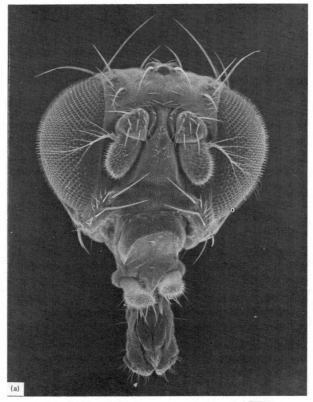

Fig. 8. Expression of homeotic mutant *antennapedia* (a) Wild-type *Drosophila* head. (b) *Antp*, with a leg substituting for much of the antenna. (Courtesy of Dr. Douglas Sears and Dr. John Postlethwait.)

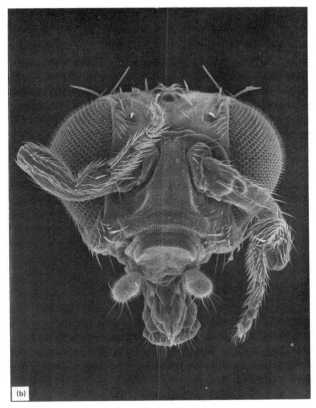

(b)

Fig. 8. *Continued.*

variety of nongenetic experimental approaches are also available for studying the phenotypic changes seen in these mutants. These include the ability to duplicate the changes in developmental specificity using explanted tissues and the ability to produce phenocopies in intact animals using a number of physical and chemical agents. This facility, together with the elegance of *Drosophila* genetics, has stimulated a great deal of study of these mutants, especially among *Drosophila* species (for reviews, see Postlethwait and Schneiderman, 1973; Oberlander, 1972; Gehring and Nöthiger, 1973; Ouweneel, 1976). The basic fascination of the system lies in the concept that mutations and experimental manipulations that alter specificity, rather than simply destroy it, are more likely to give important information about the original determination of specificity.

Antennapedia, like other homeotic mutant phenotypes, results from a change in the developmental fate of one of the pairs of imaginal discs that normally give rise to the adult integument. Imaginal discs are present to varying degrees in insects that undergo complete metamorphosis. In *Drosophila* nine pairs of imaginal discs give rise to the integument of the adult head and thorax. In the thorax, the ventral prothoracic, mesothoracic, and metathoracic discs give rise to ventral regions and the three pairs of legs. The dorsal regions are derived from three more pairs of discs, the prothoracic, mesothoracic (wing), and metathoracic (haltere) discs. The head integument, including the antennae, are derived from three more— the eye–antennal, labial, and clypeo–labrum discs. The genitalia arise from an additional single unpaired disc. The abdominal segments, rather than being formed from imaginal discs, derive from small clumps of histoblasts, and parts of the adult digestive tract and salivary glands are formed from separate imaginal rings.

At the larval stage each disc is a hollow folded sac formed of a single layer of epithelial cells (Poodry and Schneiderman, 1971). At metamorphosis, they differentiate in response to the hormonal changes occurring, among which the disappearance of juvenile hormone appears to be especially important (for reviews, see Oberlander, 1972; Doane, 1973). The product in each case is a specific set of structures whose cuticle is characterized by a regionally differentiated pattern of hairs, bristles, and sensilla.

Individual discs can be excised and implanted into the abdominal cavity of adult flies, where they will grow but do not differentiate. In this manner they can be serially transferred and propagated indefinitely. If at any time, however, they are implanted into a larval host, they respond to the hormonal changes accompanying metamorphosis by differentiating and secreting regionally specialized cuticle (Bodenstein, 1943). Prolonged passage as undifferentiated cells in adult hosts results in changes in developmental potential, so that early passages give rise to structures appropriate to the disc of origin, but later passages may differentiate into structures characteristic of other discs. This change in specificity, called transdetermination (Hadorn, 1965), has been studied extensively by Hadorn and associates (Hadorn, 1965; Gehring and Nöthiger, 1973).

In several important respects the changes seen in transdetermination are similar to those resulting from homeotic mutations. Most homeotic mutants carry a very distinct change in specificity from one set of structures to another, from one imaginal disc identity to another. Not all possible changes in specificity have been observed, and to a large measure they parallel the transdeterminative changes that have been found in culture. Indeed, there tends to be a polarity of possible changes in both cases, generally following the sequence genitalia → (antenna, leg) → wing → mesothorax. The parallelism is not complete, however.

Not all homeotic mutant changes in phenotype are observed in transdetermination experiments. It is also true that not all transdeterminative changes are represented among the known homeotic mutations, but this might be because the missing mutants are either lethal or have simply not occurred.

An interesting aspect of imaginal disc function is the ease with which phenocopies can be produced. For example, the change from antenna to leg characteristic of *antennapedia* not only results from mutation and as a transdetermination in culture (Balkaschina, 1929), but also occurs in intact flies from exposure to nitrogen mustard (Bodenstein and Abdel-Malek, 1949), X rays (Villee, 1946; Waddington, 1942), borate compounds (Gersh, 1946; Goldschmidt and Piternick, 1957; Sang and McDonald, 1954), and 5-fluorouracil (Gehring, 1964).

B. Characteristics of Homeotic Mutants

Although each homeotic mutant is unique in detail, there are several generalizations that apply to most of the homeotic genes in *Drosophila*.

The first is that most of the mutations have occurred in a few very limited regions of the genome, particularly two short segments of chromosome 3. This suggests that they are not samples of a very large number of possible metabolic or structural defects, but rather a set of genes concerned with some very restricted, and presumably related, set of processes.

The second fact is that determination, recognized as the commitment to give rise to a particular set of structures, does not occur until long after the decision to become an imaginal disc cell has occurred. The imaginal discs themselves are formed during late embryogenesis and the beginning of larval life (Laugé, 1967; Auerbach, 1936). Determination, however, does not occur until the second or third larval instar. This conclusion is reached both from studies of X ray-induced somatic crossing-over and from identification of the temperature-sensitive period of temperature-sensitive mutants. When somatic crossing-over is induced by irradiation in heterozygotes carrying wild-type and mutant alleles of a homeotic gene, sister clones of cells can arise where one is homozygous mutant and the other homozygous wild type. By asking how late in development homozygous mutant cells could be generated in this way and still express their mutant phenotype, Lewis (1964) was able to show that determination for the homeotic mutant *bithorax* (Fig. 9) does not occur until late in the third larval instar. Similar results were obtained with the mutants *aristapedia* (Roberts, 1964) and *extra sex comb, Scx* (Tokunaga and Stern, 1965). The timing of determination also has been tested in temperature-

Fig. 9. A *Drosophila* with four wings: a double homeotic mutant phenotype including *bithorax*. (From Lewis, 1963.)

sensitive homeotic mutants by asking when the temperature-sensitive
period for phenotype alteration occurs. A series of cold-sensitive alleles
of *aristapedia* have their sensitive period during the third larval instar
(Gloor and Kobel, 1966; Grigliatti and Suzuki, 1971; Villee, 1942, 1943,
1944a; Vogt, 1946) and the heat-sensitive period of the mutant *eyeless-
ophthalmoptera* allele *opt^G* occurs during the second larval instar (Postlethwait, 1973).

   The third characteristic of homeotic mutants is that the developmental
potential of mutant imaginal disc cells is probably an autonomous property of the cells. This is suggested by the results with X ray-induced somatic crossing-over. In that case small clones of mutant *aristapedia* cells
developed autonomously into a mutant phenotype, even though sur-
rounded by normal cells, when the mosaics were induced early enough
(Roberts, 1964). Imaginal disc cells have also been dissociated and then
reaggregated into new combinations whose developmental potential was
subsequently tested by implanting them into larvae. Postlethwait and
Schneiderman (1973), in a review of this literature, conclude that disc
specificity is an autonomous property of a cell but that positional specificity is not. That is, which structure a cell will attempt to give rise to is an
autonomous property of that cell, but which *part* of the structure it will
form depends upon the position of the cell in the disc and the signals it
receives from its neighbors. Thus, the genotype of a cell determines what
commitment it will make, in terms of the structure it will give rise to, but

there also appears to be an integrative environment that is generated collectively by the cells of a disc and that signals each cell as to which part of a structure it should enter.

The fourth point is that homeotic genes probably code for proteins. The principal argument for this conclusion is the occurrence of temperature-sensitive mutants of homeotic genes, such as *eyeless-ophthalmoptera* which forms a normal eye at 17°C but in which the eye cells are replaced by wing cells in flies raised at 29°C (Postlethwait, 1973). The observation of cold-sensitive mutants is a further indication that homeotic genes provide a protein function. The fact that alleles of *aristopedia* are cold sensitive has been interpreted (Postlethwait and Schneiderman, 1973) as suggesting that a change in the conformational state of the *aristopedia* gene product at low temperature is the cause of the mutant phenotype. Supporting this interpretation is the remarkable behavior of the *proboscipedia* mutation, which causes a change from a proboscis to an antennalike structure in flies raised at 15°C and to a leglike structure in flies grown at 29°C (Villee, 1944b). However, if homeotic genes provide specific proteins that somehow function in the determination of developmental specificity, it is puzzling as to why it is relatively easy to produce phenocopies of several homeotic changes using either chemicals or irradiation. It is not easy to see how X rays can inhibit a specific protein function.

The fifth, and in many ways the most provocative, fact about homeotic mutants is that their occurrence is almost exclusively limited to the imaginal disc structures of insects. Mutations causing a primordial structure to differentiate into an inappropriate product are rarely observed in other organisms, or even involving nonimaginal disc structures in *Drosophila*. Among the few occurrences of such mutants elsewhere, a very interesting group are the various mutations causing reversal of the sexual phenotype. The *testicular feminization, tfm,* mutants of both man (Morris, 1953) and mouse (Lyon and Hawkes, 1970; Ohno and Lyon, 1970) serve as a prototype of these. In each case the presence of the mutation causes a switch from differentiation of malelike to femalelike external genitalia in individuals that carry a Y chromosome, and would otherwise form male genitalia. The mechanism of the mutation is fairly well understood. It involves the fact that differentiation of secondary sexual characteristics is normally female in phenotype, unless there is an override by an androgen-stimulated process (Jost, 1953, 1970). For this override to occur in genetic males, that is animals carrying a Y chromosome, requires both the production of adequate amounts of testosterone and the presence of androgen receptor protein. The single site mutation, *tfm,* that causes a deficiency of androgen receptor protein (Bullock and Bardin, 1972; Gehring

and Tomkins, 1974; Attardi and Ohno, 1974) results in the complete substitution of female for male external genitalia in otherwise genotypic
males.

C. Models and Questions

The potential importance of homeotic mutants lies in the possibility that
an understanding of their mechanism may contribute to more general
models of developmental regulation. In attempting to assess this it may be
fruitful to consider these mutants in the context of what is known about
hormonal responsive tissues, since the differentiation of imaginal disc epithelial cells does, after all, involve a response to the hormonal changes
accompanying metamorphosis. In particular, three factors are known to
enter into the differentiative response of a tissue to a hormone. The first is
the appearance of the hormonal trigger, the second is the presence of the
necessary receptor system in the cell, and the third is the prior differentiation of the specificity factor(s) determining the ability of various genes to
respond to the stimulation. (For a more detailed discussion of some aspects of the development of specific gene responses see Section V). In
this framework it is possible to construct a reasonably plausible model of
imaginal disc differentiation that can encompass most of the known facts,
although there is no direct evidence that it is necessarily correct. The utility of doing so is that, whether or not the model is correct in detail, it does
help to focus the questions involved.

We may consider that, like sexual differentiation in mammals, the differentiation of imaginal disc cells is a hormonal response in which one developmental fate is superimposed on another. The integrative behavior of
a disc suggests that a hormone activating differentiation is produced by
the disc itself, in a cooperative manner involving a positive feedback system. That is, the appearance of a trigger, such as ecdysone, initiates the
production of a differentiation hormone whose presence, in turn, stimulates neighboring cells to produce more of the same hormone. Such a
cycle is similar to the cAMP response seen in slime molds that is the basis
of the acrasin aggregation system. To account for the hierarchy of structures that imaginal discs can give rise to requires that at least several hormones are involved. If these function in a combinatorial manner, the number of hormones required is less than the number of developmental
specificities observed. This would be true, for example, if the presence of
all of the hormonal requirements for one developmental pathway caused
that pathway to predominate over pathways with lesser requirements, as
is the case in the differentiation of mammalian genitalia. The simultaneous
presence of hormone systems A and B would impose a developmental

fate distinct from that caused by A or B alone; and these fates, in turn, would be different from each other and from the fate that would result if neither A or B was present. Indeed, formal mathematical models to account for imaginal disc differentiation have been developed based on a concept of sequential binary decisions, leading to a binary combinatorial code for each developmental fate (Garcia-Bellido, 1975; Morata and Lawrence, 1977; Kauffman *et al.*, 1978).

The important factors, then, deciding the developmental fate of an imaginal disc would be the capacity to produce hormonal signals, the presence of hormonal receptor systems, and the prior differentiation that makes certain structural genes responsive to hormonal stimulation. From this standpoint, transdetermination is most easily understood as a progressive loss of hormone responsiveness on prolonged cultivation, probably by loss of receptor protein. Homeotic mutations probably do not represent changes in hormone production, since they are cell autonomous. Rather, homeotic mutations could represent either defects in receptor protein or changes in the programming of gene responsiveness.

The use of a hormonal model serves to focus the crucial question of whether the homeotic genes are structural and code for hormonal receptors, or programmatic and represent a component of the temporal gene system determining the acquisition of hormone responsiveness by specific structural genes. (Conceivably, homeotic mutations do not affect receptor structural genes directly but indirectly through cis-acting regulatory elements attached to them. However, in effect, this is only an additional manner of identifying the mutations as affecting receptor function.) Once framed, the importance of the question as to whether homeotic genes are structural or programmatic in function becomes independent of the specific model that generated it. For if homeotic genes are truly programmatic in function, they define the existence of master controllers determining the programming of many structural genes simultaneously and make the further statement that such genes probably function through a protein product.

Unhappily, no experimental fact clearly favors assigning homeotic genes either a structural or programmatic function. It is, of course, simpler to assume that the homeotic mutations define a set of structural genes for hormonal receptors. It is also true that there is no independent evidence for the existence of master genes capable of programming the responsiveness of entire sets of structural genes. However, should such genes exist, the homeotic genes represent the most likely candidates for membership in this group.

The hormonal model has the subsidiary utility of accounting for the major features of imaginal disc determination in *Drosophila*, especially if the homeotic genes determine receptor function. The chromosomally limited location of the homeotic genes reflects the relatively small number of hormones and receptors involved. The relatively late developmental function of homeotic genes in the second and third instar suggests the appearance of the receptor system at this time. The integrative aspects of imaginal disc function derive from the cooperative production of hormone, while the developmental autonomy of mutant cells reflects their inability to respond to a given hormonal signal. The requirement that hormonal receptors undergo a conformational change to function would readily account for the existence of cold- as well as heat-sensitive mutants. Interestingly, the similar polarity of developmental changes seen among homeotic mutants, on the one hand, and after transdetermination, on the other, would reflect the fact that both describe the hierarchy of developmental possibilities defined by the set of hormonal responses originally present in a given disc.

D. Conclusions

In summary, it is the extremely orderly physical movements of cells in *Drosophila* (Hotta and Benzer, 1972), together with the sequestering of the integument forming cells in the imaginal discs, that makes the homeotic mutants of *Drosophila* appear so unique. Although it is an attractive hypothesis that these genes represent master controllers, there is no necessity to assume this. Instead, homeotic genes may code for hormone receptor systems. If so, an important feature of imaginal disc development would be the earlier act of differentiation making certain structural genes responsive to hormonal activation. Should this be the case, the significance of the homeotic mutants of *Drosophila* will turn out to lie in defining a concept of combinatorial specificity of hormone action, a somewhat different contribution to developmental biology than originally expected, but one that is perhaps as significant.

V. INTERCELLULAR SIGNALS

There is a great deal of experimental evidence that intercellular signals play an important role in guiding development. This is comforting because in virtually any view of differentiation the participation of such signals becomes a logical necessity. To the extent that differentiation results from sequential metabolic stimuli, each inducing the next, a requirement for intercellular signals is obvious. At the least, they would serve to integrate the development of different tissues. However, the same requirement exists even if differentiation is primarily driven by an intrinsic nuclear programming machinery. The organization of higher eukaryotes is so complex that a totally rigid program seems impossible. Cells must have external references to adjust to, and mechanisms must exist that maintain coordination between separate, but interacting, cell differentiation lineages.

It is useful to consider the information content of intercellular signals in assessing the possible roles that cell–cell interactions might play in directing development and the process of nuclear differentiation. This is so because the information content of a signal places strong constraints on its possible mechanism of action. The question of information content can be framed initially by examining two disparate biological processes: the induction of an enzyme protein by a steroid hormone in a responsive cell and the stimulation of an antibody response by an antigen.

In the case of the steroid, the signal is a small molecule of low information content; the biological sophistication lies in the responding system. The structural complexity of the steroid receptor protein allows it to sort out the appropriate signal from the variety of steroids present. In addition, there is a nuclear system that decides which genes will respond to the steroid–receptor complex when it enters the nucleus. For example, a mouse administered testosterone induces β-glucuronidase in kidney epithelial cells, the major urinary protein in liver, and rennin in submandibular salivary gland cells. The source of these crucial biological specificities does not lie in the structure of the testosterone molecule. Instead, it lies in the functional states of chromatin established during the course of nuclear differentiation. Thus, although on the surface it may be the appearance of a hormonal trigger that initiates a developmental change, in reality it was the prior differentiation of an appropriate state of receptivity that made the response possible. The acquisition of this responsive state includes a process of nuclear differentiation as well as the presence of a receptor system.

On the other hand, consider the process of antibody induction, where the primary cell reacting with antigen is the macrophage. Evidence has been presented that in response to an antigen a macrophage may transfer to a plasma cell an informational molecule, an RNA, carrying the sequence instructions for synthesis of an antibody directed against that antigen (Fishman and Adler, 1967, 1976; Saito et al., 1968; Cohen, 1976). Although the issue of immune RNA is still debated, the concept illustrates the point that a donor cell may be able to provide complex sequence and regulatory information to a recipient cell and thus alter its pathway of differentiation.

At the present time, we do not know with certainty whether the transfer of complex information occurs. However, it is important to recognize that information transfer does not provide an explanation for the primary driving mechanism in differentiation, any more than exposure to testosterone does. For if such transfer does occur, it only shifts the problem from one cell type to another. In order to transmit an informational RNA, a macrophage must first differentiate the capacity to produce that RNA in response to a specific antigen. If ectoderm responds to a signal from neural cells in the optic vesicle by making a lens, then those neural cells have differentiated the capacity to send the necessary information. Other neural cells are incapable of doing so. Moreover, only some ectodermal cells are capable of responding, and other morphologically similar ectodermal cells cannot. Again, as in the case of steroid hormones, the question reduces to the source of the primary programming and with it to the origins of the specialized abilities that allow cells to send, or respond to, an inductive signal.

Much frustrating effort has gone into the question of the chemical identity of the signals that pass during some of the familiar embryonic inductions, such as the organization of the amphibian embryo by the dorsal lip of the blastopore or the formation of a lens in epidermis following the arrival of the underlying optic vesicle. This has proved a technical problem of great magnitude, made even more complex by the ability of various nonspecific agents to mimic these effects. As yet, no answer is available (for review, see Saxén, 1975). It may be that the intercellular signals important in guiding development are conditional triggers, similar in their information content to steroid hormones whose signal can only vary in its intensity. On the other hand, intercellular signaling may involve informational macromolecules that can carry instructions for more complex processes. We can say, however, that in the context of nuclear differentiation and the regulatory machinery that programs it, intercellular signaling relates primarily to the maintenance of correlated development between cells and less to the underlying programming mechanisms themselves.

VI. CONCLUSIONS

Studies of the genetic controlling elements defined by McClintock provide direct evidence that at least one kind of intrinsic nuclear programming exists. Studies of temporal gene regulation of enzyme development in several organisms indicate that some form of genetically determined nuclear programming is an important mechanism regulating macromolecular differentiation of cells. Studies of homeotic mutants of *Drosophila* raise the possibility that a rather simple system of genetic switches may exist to shunt cells into one differentiated cell lineage or another. Collectively, there is a strong implication that developmental control is a relatively deterministic process, in which programming by a nuclear genetic machinery plays a large part. The developmental controls for each enzyme are relatively independent and appear to rely on a rather small number of discrete elements. The situation appears to be much simpler than was previously suggested by studies of morphological mutants. The difference probably lies in the difficulties of identifying the primary gene products affected in morphogenetic mutants and the confusion which then arises in distinguishing changes in programming from structural changes in proteins.

It is true that we remain uncertain as to whether the larger histological and morphogenetic changes in development have their own controls, or whether they represent the summation of many individual protein regulations. However, it is also true that the basic concept of a genetically determined nuclear program for differentiation remains very similar whichever is the case.

These are rather optimistic conclusions, suggesting that the processes of differentiation and development are relatively amenable to genetic analysis. It is a much more favorable situation than if differentiation is primarily regulated metabolically, where the interplay of many components must be known to understand each step and where any change has many consequences. Except for *Drosophila,* where the additional possibility of examining homeotic mutants exists, the key at the present time appears to be in directing our analysis at the study of individual gene products. In much the same sense that the principles describing molecular genetics and the determination of protein structure are independent of the metabolic functions an enzyme performs, it may be possible to understand the rules of developmental programming before we fully understand the complex physiological processes of morphogenesis.

REFERENCES

Abe, K. (1958). *Jpn. J. Genet.* **33**, 138–145.
Abraham, I., and Doane, W. W. (1976). *Genetics* **83**, 51.
Abraham, I., and Doane, W. W. (1978). *Proc. Natl. Acad. Sci. U. S. A.* **75**, 4446–4450.
Attardi, B., and Ohno, S. (1974). *Cell* **2**, 205–212.
Auerbach, C. (1936). *Trans. R. Soc. Edinburgh* **58**, 787–815.
Bahn, E. (1968). *Hereditas* **58**, 1–12.
Bahn, E. (1971). *Hereditas* **67**, 75–78.
Balkaschina, E. I. (1929). *Wilhelm Roux' Arch. Entwicklungsmech. Org.* **115**, 448–463.
Berger, F., Paigen, K., and Meisler, M. (1978). *J. Biol. Chem.* **253**, 5280–5282.
Bodenstein, D. (1943). *Biol. Bull. (Woods Hole, Mass.)* **84**, 34–58.

58   K. PAIGEN

Bodenstein, D., and Abdel-Malek, A. (1949). *J. Exp. Zool.* **111,** 95–114.
Boubelik, M., and Lengerova, A. (1975) *Folia Biol. (Prague)* **21,** 81.
Boubelik, M., Lengerova, A., Bailey, D. W., and Matousek, V. (1975). *Dev. Biol.* **47,** 206–214.
Brawerman, G. (1976). *Cancer Res.* **36,** 4278–4281.
Breen, G., Lusis, A. J., and Paigen, K. (1977). *Genetics* **85,** 73–84.
Britten, R. J., and Davidson, E. H. (1969). *Science* **165,** 349–357.
Bullock, L. P., and Bardin, C. W. (1972). *J. Clin. Endocrinol. Metab.* **35,** 935–937.
Childs, B., and Young, W. J. (1963). *Am. J. Med.* **34,** 663.
Chovnick, A., Gelbart, W., McCarron, M., and Osmond, B. (1976). *Genetics* **84,** 233–255.
Chovnick, A., Gelbart, W., and McCarron, M. (1977). *Cell* **11,** 1–10.
Cohen, E. P. (1976). "Immune RNA." CRC Press, Cleveland, Ohio.
Coleman, D. L. (1966). *J. Biol. Chem.* **241,** 5511–5517.
Courtright, J. B. (1967). *Genetics* **57,** 25–39.
Dagg, C. P., Coleman, D. L., and Fraser, G. M. (1964). *Genetics* **49,** 979–989.
Daniel, W. L. (1976a). *Biochem. Genet.* **14,** 1003–1018.
Daniel, W. L. (1976b). *Genetics* **82,** 477–491.
Dickinson, W. J. (1970). *Genetics* **66,** 487–496.
Dickinson, W. J. (1971). *Dev. Biol.* **26,** 77–86.
Dickinson, W. J. (1972). *Genetics* **71,** 514.
Dickinson, W. J. (1975). *Dev. Biol.* **42,** 131–140.
Dickinson, W. J., and Sullivan, D. T. (1975). "Gene-Enzyme Systems in Drosophila." Springer-Verlag, Berlin and New York.
Dizik, M., and Elliott, R. W. (1977). *Biochem. Genet.* **15,** 31–46.
Dizik, M., and Elliott, R. W. (1978). *Biochem. Genet.* **16,** 247–260.
Doane, W. W. (1965). *Am. Zool.* **5,** 697.
Doane, W. W. (1967a). *J. Exp. Zool.* **164,** 363–378.
Doane, W. W. (1967b). *Am. Zool.* **7,** 780.
Doane, W. W. (1969a). *J. Exp. Zool.* **171,** 321–342.
Doane, W. W. (1969b). *In* "Problems in Biology: RNA in Development" (E. W. Hanley, ed.), pp. 73–109. Univ. of Utah Press, Salt Lake City.
Doane, W. W. (1971). *Isozyme Bull.* **4,** 46–48.
Doane, W. W. (1973). *In* "Developmental Systems: Insects" (S. J. Counce and C. H. Waddington, eds.), Vol. 2, pp. 291–498. Academic Press, New York.
Doane, W. W. (1977). *Genetics* **86,** s15–16.
Doane, W. W., Abraham, I., Kolar, M. M., Martenson, R. E., and Deibler, G. E. (1975). *In* "Isozymes" (C. L. Markert, ed.), Vol. 4, pp. 585–607. Academic Press, New York.
Doyle, D., and Schimke, R. T. (1969). *J. Biol. Chem.* **244,** 5449–5459.
Endo, T., and Schwartz, D. (1966). *Genetics* **54,** 233–239.
Felton, J., Meisler, M., and Paigen, K. (1974). *J. Biol. Chem.* **249,** 3267–3272.
Fincham, J. R. S. (1973). *Genetics* **73,** 195–205.
Fincham, J. R. S., and Sastry, G. R. K. (1974). *Annu. Rev. Genet.* **8,** 15–50.
Fishman, M., and Adler, F. L. (1967). *Cold Spring Harbor Symp. Quant. Biol.* **32,** 343–348.
Fishman, M., and Adler, F. L. (1976). *In* "Immune RNA in Neoplasia" (M. A. Fink, ed.), pp. 53–60, Academic Press, New York.
Ganschow, R. E. (1975). *In* "Isozymes" (C. L. Markert, ed.), Vol. 4, pp. 633–647. Academic Press, New York.
Ganschow, R. E., and Bunker, B. G. (1970). *Biochem. Genet.* **4,** 127–133.
Ganschow, R. E., and Paigen, K. (1967). *Proc. Natl. Acad. Sci. U. S. A.* **58,** 938–945.
Ganschow, R. E., and Schimke, R. T. (1969). *J. Biol. Chem.* **244,** 4649–4658.
Garcia-Bellido, A. (1975). *Ciba Found. Symp.* **29** (New Ser.), 161.

Gehring, U., and Tomkins, G. M. (1974). *Cell* **3**, 59–64.
Gehring, W. J. (1964). *Drosophila Inf. Serv.* **39**, 102.
Gehring, W. J. and Nöthiger, R. (1973). *In* "Developmental Systems: Insects" (S. J. Counce and C. H. Waddington, eds.), Vol. 2, pp. 212–290. Academic Press, New York.
Gerdes, R. A. (1975). *Genetics* **80**, s34.
Gersh, E. (1946). *Drosophila Inf. Serv.* **20**, 86.
Gierer, A. (1973). *Cold Spring Harbor Symp. Quant. Biol.* **38**, 951–961.
Glassman, E., Shinoda, T., Duke, E. J., and Collins, J. F. (1968). *Ann. N. Y. Acad. Sci.* **151**, 263–273.
Gloor, H., and Kobel, H. (1966). *Rev. Suisse Zool.* **73**, 229–225.
Goldschmidt, R., and Piternick, L. (1957). *J. Exp. Zool.* **135**, 127–202.
Grey, H. M., Kubo, R. T., Colon, S. M., Poulik, M. D., Cresswell, P., Springer, T. A., Turner, M., and Strominger, J. L. (1973). *J. Exp. Med.* **131**, 1608–1612.
Grigliatti, T., and Suzuki, D. T. (1971). *Proc. Natl. Acad. Sci. U. S. A.* **68**, 1307–1311.
Gusella, J., Geller, R., Clarke, B., Weeks, V., and Housman, D. (1976). *Cell* **9**, 221–229.
Hadorn, E. (1965). *Brookhaven Symp. Biol.* **18**, 148–161.
Harris, H. (1964). *In* "Congenital Malformations" (M. Fishbein, ed.), pp. 135–144. Int. Med. Congr., Ltd., New York.
Herrup, K., and Mullen, R. J. (1977). *Biochem. Genet.* **15**, 641–653.
Holliday, R., Huschtscha, L. I., Tarrant, G. M., and Kirwood, T. B. L. (1977). *Science* **198**, 366–372.
Hotta, Y., and Benzer, S. (1972). *Nature (London)* **240**, 527–535.
Johnson, F. M. (1966). *Nature (London)* **212**, 843–844.
Jost, A. (1953). *Recent Prog. Horm. Res.* **8**, 379–418.
Jost, A. (1970). *Philos. Trans. R. Soc. London* **259**, 119.
Kaplan, R. D., Chapman, V., and Ruddle, F. (1973). *J. Hered.* **64**, 155–157.
Karl, T. R., and Chapman, V. M. (1974). *Biochem. Genet.* **5**, 367–372.
Kauffman, S. A., Shymko, R. M., and Trabert, K. (1978). *Science* **199**, 259–270.
Kikkawa, H. (1960). *Jpn. J. Genet.* **33**, 382–387.
Kikkawa, H. (1964). *Jpn. J. Genet.* **39**, 401–411.
King, J. C. (1969). *Proc. Natl. Acad. Sci. U. S. A.* **64**, 891–896.
Kint, J. A. (1970). *Science* **167**, 1268–1269.
Klein, J. (1975). "Biology of the Mouse Histocompatibility-2 Complex." Springer-Verlag, Berlin and New York.
Kozak, C., Nichols, E., and Ruddle, F. H. (1975). *Somatic Cell Genet.* **1**, 371–382.
Lalley, P. A., and Shows, T. B. (1974). *Science* **185**, 442–444.
Lalley, P. A., and Shows, T. B. (1977). *Genetics* **87**, 305–317.
Laugé, G. (1967). *C. R. Hebd. Seances Acad. Sci.* **265**, 814–817.
Law, L. W., Morrow, A. G., and Greenspan, E. M. (1952). *J. Natl. Cancer Inst.* **12**, 909–916.
Lefevre, G. (1974). *Annu. Rev. Genet.* **8**, 51–62.
Lewis, E. B. (1963). *Am. Zool.* **3**, 33–56.
Lewis, E. B. (1964). *In* "The Role of Chromosomes in Development" (M. Locke, ed.), pp. 231–252. Academic Press, New York.
Lindsley, D. L., and Grell, E. H. (1967). *Carnegie Inst. Washington Publ.* **627**, 59.
Lundin, L.-G., and Seyedyazdani, R. (1973). *Biochem. Genet.* **10**, 351–361.
Lusis, A. J., and Paigen, K. (1975). *Cell* **6**, 371–378.
Lusis, A. J., and Paigen, K. (1976). *Biochim. Biophys. Acta* **437**, 487–497.
Lusis, A. J., and Paigen, K. (1977). *In* "Isozymes" (M. C. Rattazzi, J. G. Scandalios, and G. S. Whitt, eds.), Vol. 2, pp. 63–106. Alan R. Liss, Inc., New York.
Lusis, A. J., and West, J. D. (1976). *Biochem. Genet.* **14**, 849–855.

Lusis, A. J., and West, J. D. (1978). *Genetics* **88**, 327–342.
Lusis, A. J., Tomino, S., and Paigen, K. (1977). *Biochem. Genet.* **15**, 115–122.
Lyon, M. F., and Hawkes, S. G. (1970). *Nature (London)* **227**, 1217–1219.
McClintock, B. (1951). *Cold Spring Harbor Symp. Quant. Biol.* **16**, 13–47.
McClintock, B. (1965). *Brookhaven Symp. Biol.* **18**, 162–184.
McClintock, B. (1967). *Dev. Biol.* **1**, 84–112.
Meisler, M. H. (1976). *Biochem. Genet.* **14**, 921–932.
Meisler, M. H., and Paigen, K. (1972). *Science* **177**, 894–896.
Morata, G., and Lawrence, P. A. (1977). *Nature (London)* **265**, 211–216.
Morris, J. M. (1953). *Am. J. Obstet. Gynecol.* **65**, 1192.
Nakamuro, K., Tanigaki, M., and Pressman, D. (1973). *Proc. Natl. Acad. Sci. U. S. A.* **70**, 2863–2865.
Nielsen, J. T., and Sick, K. (1964). *Hereditas* **51**, 291–296.
Noell, W. K. (1958). *AMA Arch. Ophthalmol.* **60**, 702.
Oberlander, H. (1972). *Results Probl. Cell Differ.* **5**, 155–172.
Ohno, S., and Lyon, M. F. (1970). *Clin. Genet.* **1**, 121–127.
Olsson, M., Lindahl, G., and Ruoslahti, E. (1977). *J. Exp. Med.* **145**, 819–827.
Ouweneel, W. J. (1976). *Adv. Genet.* **18**, 179–248.
Paigen, K. (1961a). *Proc. Natl. Acad. Sci. U. S. A.* **47**, 1641–1649.
Paigen, K. (1961b). *Exp. Cell Res.* **25**, 286–301.
Paigen, K. (1964). *In* "Congenital Malformations" (M. Fishbein, ed.), pp. 181–190. Int. Med. Congr., Ltd., New York.
Paigen, K. (1971). *In* "Enzyme Synthesis and Degradation in Mammalian Systems" (M. Rechcigl, ed.), pp. 1–47. Univ. Park Press, Baltimore, Maryland.
Paigen, K. (1977). *Proc. Int. Congr. Hum. Genet., 5th, 1976* Excerpta Med. Int. Congr. Ser. No. 411, pp. 33–42.
Paigen, K., and Ganschow, R. (1965). *Brookhaven Symp. Biol.* **18**, 99–115.
Paigen, K., Swank, R. T., Tomino, S., and Ganschow, R. E. (1975). *J. Cell. Physiol.* **85**, 379–392.
Paigen, K., Meisler, M., Felton, J., and Chapman, V. (1976). *Cell* **9**, 533–539.
Peterson, P. A., Rask, L., and Lindblom, J. B. (1974). *Proc. Natl. Acad. Sci. U. S. A.* **71**, 35–39.
Poodry, C. A., and Schneiderman, H. A. (1971). *Wilhelm Roux' Arch. Entwicklungsmech. Org.* **168**, 1–9.
Postlethwait, J. H. (1973). Quoted in Postlethwait and Schneiderman (1973, p. 428).
Postlethwait, J. H., and Schneiderman, H. A. (1973). *Annu. Rev. Genet.* **7**, 381–434.
Rawls, J. M., and Fristrom, J. W. (1975). *Nature (London)* **255**, 738–740.
Rebourcet, R., Weil, D., Van Cong, N., and Frezal, J. (1975). *Cytogenet. Cell Genet.* **14**, 406–408.
Rechcigl, M., and Heston, W. E. (1967). *Biochem. Biophys. Res. Commun.* **27**, 119–124.
Roberts, P. (1964). *Genetics* **49**, 593–598.
Roy, A. B. (1958). *Biochem. J.* **68**, 519.
Saito, K., Kurashige, S., and Mitsuhashi, S. (1968). *Jpn. J. Microbiol.* **13**, 122–124.
Sang, J., and McDonald, J. (1954). *J. Genet.* **52**, 392–412.
Sanno, Y., Holzer, M., and Schimke, R. T. (1970). *J. Biol. Chem.* **245**, 5668–5676.
Saxén, L. (1975). *Clin. Obstet. Gynecol.* **18**, 149–175.
Schwartz, D. (1960). *Proc. Natl. Acad. Sci. U. S. A.* **46**, 1210–1215.
Schwartz, D. (1962). *Genetics* **47**, 1609–1615.
Schwartz, D. (1964a). *Natl. Cancer Inst., Monogr.* **18**, 9–14.
Schwartz, D. (1964b). *Proc. Natl. Acad. Sci. U. S. A.* **52**, 222–226.
Schwartz, D. (1964c). *Genetics* **49**, 373–377.

Schwartz, D. (1964d). *In* "Structure and Function of the Genetic Material," pp. 201–203. Akademie-Verlag, Berlin.

Schwartz, D. (1965). *Genetics* **52,** 1295–1302.

Schwartz, D. (1967). *Proc. Natl. Acad. Sci. U. S. A.* **58,** 568–575.

Schwartz, D., Fuchsman, L., and McGrath, K. H. (1965). *Genetics* **52,** 1265–1268.

Silver, J., and Hood, L. (1974). *Nature (London)* **249,** 764–765.

Stern, R. H., Russell, E. S., and Taylor, B. A. (1976). *Biochem. Genet.* **14,** 373–381.

Suzuki, K., and Suzuki, K. (1973). *In* "Lysosomes and Storage Diseases" (H. G. Hers and F. van Hoof, eds.), pp. 395–410. Academic Press, New York.

Suzuki, Y., Gage, P., and Brown, D. D. (1972). *J. Mol. Biol.* **70,** 637–649.

Swank, R. T., and Bailey, D. W. (1973). *Science* **181,** 1249.

Swank, R. T., and Paigen, K. (1973). *J. Mol. Biol.* **77,** 371–389.

Swank, R. T., Paigen, K., and Ganschow, R. (1973). *J. Mol. Biol.* **81,** 225–243.

Swank, R. T., Paigen, K., Davey, R., Chapman, V., Labarca, C., Watson, G., Ganschow, R., Brandt, E. J., and Novak, E. (1978). *Recent Prog. Horm. Res.* **34,** 401–436.

Terhorst, C., Robb, R., Jones, C., and Strominger, J. L. (1977). *Proc. Natl. Acad. Sci. U. S. A.* **74,** 4002–4006.

Tokunaga, C., and Stern, C. (1965). *Dev. Biol.* **11,** 64–82.

Tomino, S., and Meisler, M. (1975). *J. Biol. Chem.* **250,** 7752–7758.

Tomino, S., and Paigen, K. (1975). *J. Biol. Chem.* **250,** 1146–1148.

Tomino, S., Paigen, K., Tulsiani, D. R. P., and Touster, O. (1975). *J. Biol. Chem.* **250,** 8503–8509.

Van Blerkom, J., and Manes, C. (1977). *In* "Concepts in Mammalian Embryogenesis" (M. I. Sherman, ed.), pp. 37–94. MIT Press, Cambridge, Massachusetts.

Villee, C. (1942). *Am. Nat.* **76,** 494–506.

Villee, C. (1943). *J. Exp. Zool.* **93,** 75–98.

Villee, C. (1944a). *J. Elisha Mitchell Sci. Soc.* **60,** 141–157.

Villee, C. (1944b). *J. Exp. Zool.* **96,** 85–102.

Villee, C. (1946). *J. Exp. Zool.* **107,** 261–280.

Vogt, M. (1946). *Biol. Zentralbl.* **65,** 238–254.

Waddington, C. H. (1942). *J. Exp. Biol.* **19,** 101–117.

Weiss, P. A. (1973). *Differentiation* **1,** 3–10.

Wenger, D. A., Sattler, M., and Hiatt, W. (1974). *Proc. Natl. Acad. Sci. U. S. A.* **71,** 854–857.

Wudl, L., and Chapman, V. (1976). *Dev. Biol.* **48,** 104–109.

2

Control of Gene Expression and Enzyme Differentiation

JOHN G. SCANDALIOS

PHYSIOLOGICAL GENETICS
Copyright © 1979 by Academic Press, Inc.

I. INTRODUCTION

The programmed and precise expression of genes during development involves some of the most complex sequences of biochemical reactions observed in living cells. The molecular mechanisms by which information encoded in genes is decoded and translated into proteins are fairly well understood. However, the mechanism(s) whereby the cell controls or modulates the activity of its structural genes remains to be resolved. Although some of the mechanisms regulating gene expression have been resolved in prokaryotes, there is very little information on such mechanisms in the more complex eukaryotes. In fact, the data accumulated to date suggest that there are distinct and important differences in the controls identified in microbes and higher organisms.

The coordinated expression of genes observed during the development of complex eukaryotes is likely the result of the evolution of complicated regulatory mechanisms along with the increase and specialization of the genetic information as organisms evolved from the relatively simple to the more complex forms. There are, in fact, three general levels of control currently recognized: (a) "transcriptional control" (pertaining to the synthesis of messenger RNA), (b) "translational control" (pertaining to processing and utilization of messenger RNA), and (c) "posttranslational control" (pertaining to processing and modification of protein molecules following polypeptide synthesis). Each of these levels of control incorporates a number of precise points at which control can be exerted in regulating the expression of genes (Table I). Very little is currently known as to whether any one of the specific control points may be the most important in generating the ultimate phenotype. In the more complex eukaryotes these mechanisms, or control points, may operate at two basic levels of gene regulation during development and differentiation; these are (a) "spatial control" and (b) "temporal control."

Analysis of enzyme expression provides a reasonable and promising approach toward understanding the regulation of gene expression in eukaryotes, providing we accept the underlying assumption that the characteristics of a given cell at different stages of development are functions of the protein (enzyme) molecules existing in those cells. Alternatively stated, since enzymes are necessary for the chemical reactions of living cells, the molecular bases for eukaryote development may be critically analyzed by studying the ontogeny of enzymes and the mechanisms regulating their expression and function in time (temporal) and in specific tissues (spatial). Since protein synthesis is a central process in the metabolism of the cell, it directly expresses the information encoded in the genes. However, biochemical development cannot only be described in terms of

TABLE I **Some Possible Points of Control of Gene Expression**

I. Transcriptional control
 1. Availability of the DNA template for transcription
 2. Initiation of transcription (template recognition and binding of RNA polymerase to the template)
 3. Rate of transcription (number and activity of RNA polymerase molecules)
 4. Termination of transcription and release of mRNA
II. Translational control
 1. Processing and "maturation" of mRNA versus degradation (i.e., availability of mRNA for translation)
 2. Transport of mRNA
 3. Availability of mRNA and aminoacyl-tRNA synthetases
 4. Availability of ribosome subunits
 5. Formation of initiation complex
 6. Initiation
 7. Peptide bond formation and translocation
 8. Termination of protein synthesis and release of protein
 9. Continued availability of mRNA versus degradation
III. Posttranslational control
 1. Modification of protein structure (primary, secondary, tertiary, quaternary)
 2. Activation or inactivation of protein
 3. Synthesis versus degradation (rate of turnover)
 4. Intracellular processing (compartmentation)

enzyme differentiation; nonenzymatic functional and structural proteins and other molecules play an integral part in the overall developmental profile of the organism. The uniqueness of enzymes is that they are all proteins and their behavior may reflect general principles underlying the developmental characteristics of nonenzymatic proteins as well, and the fact that, as a consequence of their catalytic properties, they may reflect the metabolic and differentiated state of cells. To a geneticist, enzymes serve as sensitive probes to monitor the developmental pattern of the encoding genes in an attempt to elucidate, at least at the posttranslational level, those control mechanisms that may underlie the differential expression of genes in the complex eukaryotes.

The important phenotypic characteristics of an enzyme are its properties as a catalyst, the regulation of its synthesis and degradation, and its distribution within cells and tissues. Since the catalytic properties of an enzyme are ultimately a consequence of its amino acid sequence, they can be altered by mutation of its structural gene. Such properties include catalytic efficiency, substrate specificity, physical stability, and response to regulatory effectors.

It is now a well known and accepted fact that enzymes can exist in multiple molecular forms, or "isozymes," within the cells of a single organism. In fact, several hundred different enzymes are now known to exist in multiple molecular forms in a great variety of organisms. The subject has been discussed in detail elsewhere (Shaw, 1969; Scandalios, 1969, 1974; Eppenberger, 1975; Markert, 1968, 1977). This general phenomenon affords an opportunity for the study of the sequential development of organisms, that is, the isozymic profile of a cell reflects its state of differentiation, and, consequently, an understanding of the changing isozymic profile and the factors affecting it during development may lead to some insight as to the mechanisms involved in cellular differentiation, and more specifically of differential gene expression.

The catalase gene–enzyme system in *Zea mays* L. represents an ideal model system for the exploration of factors controlling gene expression during the development of a complex eukaryote. For this reason it will be discussed in some detail in this chapter.

II. ISOZYMES

A. Definition

The term isozyme was coined by Markert and Møller (1959) to refer to the occurrence of multiple forms of a given enzyme within the same organism. Isozymes have similar or identical substrate specificities. Enzymes with very broad substrate specificities were excluded from the initial definition of the term. Although the phenomenon of enzyme multiplicity had been known in a few instances for a long time, it was deemed by the biochemists as being indicative of artifacts in their preparations rather than suggestive of biological complexity. Acceptance of the existence of isozymes as an underlying basic biological phenomenon indicative of molecular diversity within cells and organisms did not occur until quite recently (see review by Markert, 1977). The term isozyme was intended to fill the need for a precise description of the existence of multiple forms of a given enzyme, even though the nature of such multiplicity may not at first be known (Markert, 1968). Thus the word isozyme was intended to have only operational utility; it will be used in that context throughout this chapter.

B. Molecular Bases of Isozyme Formation

Occurrence of isozymes may come about as a consequence of either or both of two general mechanisms.

1. Epigenetic mechanisms operating at the translational or posttranslational levels to modify polypeptides to varying degrees (e.g., conformational alterations, covalent modification, deamidation, adenylation, phosphorylation, selective degradation, selective or partial cleavage).

2. Genetic mechanisms operating at the genome level which are subsequently transcribed onto mRNAs and then code for different polypeptides. The polypeptides may be the final functional or structural units, or they may serve as subunits which assemble to generate functional multimers. Multimers may be composed of identical (homo) or nonidentical (hetero) subunits.

Isozymes may be encoded in allelic or nonallelic genes. They may be generated as a consequence of gene duplications with subsequent mutations at "daughter" and "parental" loci, as exemplified by the malate dehydrogenase genes in maize (Yang *et al.*, 1977). Thus, more than one gene contributes to the structure of any enzyme composed of more than one kind of subunit. In fact, a useful formula can be set up which can relate the number of isozymes to the multimer size and the number of kinds of subunits (Fig. 1). Once the molecular nature for the multiple forms of a given enzyme is understood, the term isozyme can then be modified by such adjectives as allelic, nonallelic, multigenic, homomultimeric, heteromultimeric, conformational, hybrid, and conjugated; this will more precisely reflect the degree of our knowledge of the particular isozyme system (Markert, 1968; Shaw, 1969; Scandalios, 1974).

$$I = \frac{(S + N - 1)!}{N! \, (S - 1)!}$$

	S			
	1	2	3	4
2	1	3	6	10
N 3	1	4	10	20
4	1	5	15	35

I = number of isozymes
N = number of subunits per polymer
S = number of different subunits

Fig. 1. Chart and formula for assessing subunit composition of isozyme systems. The number of observed isozymes is given in the boxes; these may be produced by combinations of different numbers (N) of distinct kinds of subunits (S). The formula allows the computation of the number of isozymes (I) that can be generated by random combination of any number of distinctly different subunits into polymers (multimers) of any size. (From Markert, 1977, with permission.)

C. Occurrence

Isozymes are now known to occur commonly among animals, plants, and microbes. Several hundred enzymes examined in a wide variety of organisms have been shown to have isozymic forms (Markert, 1975; Scandalios, 1974). Multiplicity of enzymes seems to be the rule rather than the exception.

III. THE ENZYME CATALASE: GENERAL ASPECTS

A. Some Characteristics

Catalase (H_2O_2:H_2O_2 oxidoreductase; EC 1.11.1.6) is one of the earliest enzymes to have been isolated (from beef liver extracts), purified, and characterized. The enzyme has been isolated and characterized from a variety of sources; in all cases, catalase was found to consist of four tetrahedrally arranged polypeptide subunits, each associated with a single prosthetic group, the ferric protoporphyrin IX (Schonbaum and Chance, 1976).

Catalase is one of the most efficient catalysts produced by nature. It decomposes hydrogen peroxide at an extremely rapid rate, corresponding to a catalytic center activity of about 10^7 min^{-1}. Catalase exerts a dual function, depending upon the concentration of H_2O_2 available. If the steady-state concentration of H_2O_2 in the system is high the enzyme acts *catalatically* whereby 2 moles of water and 1 mole of molecular oxygen are produced by the oxidoreduction of 2 moles of hydrogen peroxide. However, at low hydrogen peroxide concentrations ($<10^{-6}$ M) and in the presence of a suitable hydrogen donor (e.g., methanol, ethanol, phenols, primary amines), catalase acts *peroxidatically* whereby the H_2O_2 is oxidized (Aebi and Sutter, 1971). These reactions are illustrated as follows.

Catalatic reaction
$$H_2O_2 + H_2O_2 \rightarrow 2H_2O + O_2$$

where H_2O_2 is the substrate and hydrogen donor

Peroxidatic reaction
$$RH_2 + H_2O_2 \rightarrow R + 2H_2O$$

where RH_2 is the hydrogen donor (e.g., ethanol, methanol, formic acid). Detailed studies by Chance (1949) led to the conclusion that the catalatic reaction took place in two steps.

$$Cat—Fe^{3+}—OH + H_2O_2 \rightarrow Cat—Fe^{3+}—OOH + H_2O \qquad (1)$$
$$(I)$$

$$\text{Cat—Fe}^{3+}\text{—OOH} + H_2O_2 \rightarrow \text{Cat—Fe}^{3+}\text{—OH} + H_2O + O_2 \uparrow \qquad (2)$$
$$\text{(II)}$$

It is catalase compound I with an oxidizing equivalent of two which can be reduced by other hydrogen donors (e.g., ethanol) peroxidatically.

$$\text{Cat—Fe}^{3+}\text{—OOH} + C_2H_5OH \rightarrow \text{Cat—Fe}^{3+}\text{—OH} + H_2O + CH_3CHO \qquad (3)$$
$$\text{(III)}$$

Compounds I and II have recently been observed *in vivo* in hemoglobin-free perfused rat liver (Oshino and Chance, 1973a,b).

Catalase is known to be inhibited by the usual heme-binding ligands (e.g., cyanide, azide, and carbon monoxide). However, the most interesting inhibitor of catalase is the herbicide 3-amino-1,2,4-triazole (AT). Margoliash *et al.* (1960) demonstrated that this irreversible inhibition is caused by AT binding to the apoprotein of the catalase at a histidine residue near the "heme pocket" on the peptide chain. The inhibition reaction occurs by interaction between compound I and AT in the presence of H_2O_2. The inhibition by AT does not interfere with the subsequent renewal of newly synthesized catalase. In mice, another compound, the porphyrinogenic drug allylisopropylacetamide (AIA), has been shown to decrease catalase activity (Price *et al.*, 1962). However, AIA acts by blocking the synthesis of the new enzyme without interfering with the activity of previously formed catalase. Interestingly, both AT and AIA have been used in studies of catalase turnover in a variety of organisms (Rechcigl, 1971; Sorenson *et al.*, 1977).

B. Occurrence

Catalase is ubiquitously present in all aerobic cells (plants, animals, and microbes) containing a cytochrome system; only strict anaerobes seem to lack catalase activity (Singer, 1971). The enzyme is found in most tissues of the various organisms examined. In mammalian tissues there is considerable variation, with catalase concentration being highest in liver and erythrocytes and lowest in connective tissue. The liver catalase is primarily localized in peroxisomes (de Duve and Baudhuin, 1966). In a number of marine invertebrates examined, catalase activity was found to be highest in gonadal tissues and eggs (Nelson and Scandalios, 1977).

C. Genetic Variation

Genetic variants of catalase activity have been reported in numerous organisms (Feinstein, 1970) beginning with a series of little known papers by Russian investigators published during the period 1922–1927 where a number of species were shown to have blood catalase levels under genetic control (Koltzoff, 1927; Stern, 1968). However, it was the discovery by Takahara (1952) that certain human individuals lacked catalase activity (acatalasemia) in their blood, which prompted attention to the genetics of the enzyme. Takahara's pioneering studies demonstrated that acatalasemia is compatible with life. It should, however, be pointed out that the "acatalasemic" human subjects always have trace amounts of catalase in their erythrocytes (Aebi and Sutter, 1971); this amount may be sufficient to carry out certain essential metabolic reactions.

D. Biological Function and Intracellular Localization

The physiological role of catalase is far from clear. It has been suggested (de Duve and Baudhuin, 1966) that catalase normally acts peroxidatically in its physiological role with the catalatic activity serving as a "safety valve" to prevent the accumulation of toxic levels of hydrogen peroxide. In addition, these investigators have proposed that catalase may participate in a coupled "shunt" reaction with cytosolic alcohol dehydrogenase to reoxidize cytosol NADH.

Catalase has been implicated in a number of fundamental reactions which it can efficiently catalyze under varied conditions; some of these reactions may be vital to the survival of the organism in which they occur. The discovery of mutants virtually lacking catalase activity without causing any significant pathological symptoms in humans (Takahara and Miyamoto, 1948; Aebi and Sutter, 1971) or in mice (Feinstein, 1970) led to the reference of catalase as a "fossil enzyme" (Nichols and Schonbaum, 1963). The most generally acknowledged role for catalase is that it is involved in the protective destruction of hydrogen peroxide produced during cellular metabolism as a consequence of the dismutation of superoxide radicals by superoxide dismutase (McCord, 1979), various amino acid oxidases, or as a result from such exogenous factors as ionizing radiations (Freese et al., 1967). This notion is strongly supported by a large volume of experimental data, especially by the fact that catalase has clearly been shown to be associated, as a major component, with microbodies from all sources examined (de Duve and Baudhuin, 1966; Beevers, 1969; Tolbert, 1971). Two types of microbodies have been identified based on their enzyme complement, as a result of recent biochemical studies (Tolbert, 1971). The "peroxisomes" which are found in animal cells and leaves of higher plants, and "glyoxysomes" which have been described only in plants (Beevers, 1969) where they are particularly abundant in germinating seeds which store fats as reserve material. Although the two organelles contain basically different sets of enzymes, they are both characterized by the presence of flavin-linked oxidases, which produce hydrogen peroxide with the uptake of molecular oxygen, and excessive levels of catalase, which is able to break down the H_2O_2 to water and molecular oxygen. In plants, peroxisomes are involved in the synthesis of glycine and serine and probably of C-1 donor from the primary products of photosynthesis (Bird *et al.*, 1972). The glyoxysomes are known to function in the conversion of fat to sugars via the glyoxylate cycle (Cooper and Beevers, 1969).

Recently, catalase has been suggested to be involved in the promotion of dormant seed germination. Hendricks and Taylorson (1975) have postulated a reaction network whereby germination of some dormant seeds (i.e., lettuce and pigweed) is promoted as a consequence of catalase inhibition which allows metabolically derived hydrogen peroxide to oxidize reduced NADPH, which in turn serves as the required oxidant in the pentose pathway of glucose utilization.

IV. THE CATALASE GENE–ENZYME SYSTEM OF MAIZE: A MODEL SYSTEM

A. Organism of Choice

Among the higher eukaryotes, three organisms lend themselves best to physiological genetic studies. They are the invertebrate *Drosophila melanogaster,* the vertebrate *Mus musculus,* and the higher plant *Zea mays* L. or maize. In the studies I will discuss in this chapter, based primarily on research in my own laboratory, maize is the organism of choice due to the ready availability of well-characterized genetic stocks, our fairly detailed knowledge of its genetics and cytology, its well-delineated developmental stages and availability of experimental material. In addition, the seed of higher plants has many advantages in studies on protein synthesis and degradation, differential gene function (as many proteins are tissue or organ specific), and in developmental studies (since after a metabolically very active phase, the relatively metabolically inactive mature seed maintains the developmental potential to allow seed germination to proceed). Thus, seed development may be viewed as a preparatory phase for subsequent successful germination and sporophytic development. So that the reader may be properly oriented with respect to the ensuing discussion of this system, several aspects of the maize developmental morphology are summarized below (Figs. 2 and 3).

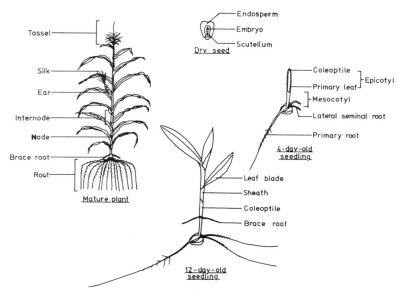

Fig. 2. Diagram of various developmental stages in the life cycle of the maize plant, *Zea mays* L.

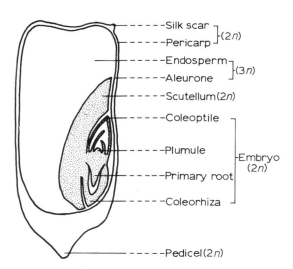

Fig. 3. Diagram of longitudinal section of a maize kernel. The ploidy condition of the various tissues is indicated since it pertains to gene dosage effects discussed in the text.

B. Genetic Control of the CAT-1 Catalases

On screening a large number of inbred maize strains from various geographic sources, it was shown that there exists one electrophoretically distinct form of catalase in the endosperm of immature kernels; six electrophoretic variants of the basic molecular form of catalase have been found to date, each characteristic of a particular inbred strain (Scandalios, 1968). It was further demonstrated that the endosperm catalase variants are controlled by six allelic genes (Scandalios, 1968) at the *Cat1* gene locus (Fig. 4a). In F_1 heterozygotes between any two variant strains, the catalase from immature (milky) endosperm is resolvable into five electrophoretically distinct forms, the two parental types and three new isozymes with intermediate electrophoretic mobilities (Fig. 4b). Since endosperm is a triploid tissue with twice the maternal genomic contribution, two kinds of heterozygous individuals are observed depending on the direction of the genetic cross (Fig. 4b). The apparent gene dosage offers an opportunity to examine how it may affect the enzyme's structure and function, and to determine the time of expression of parental alleles in heterozygotes following fertilization. The *Cat1* gene is also expressed in the scutellum (a diploid tissue) during the same developmental period, but the activity distribution is as expected of diploid tissues based on random combinations of subunits (i.e., $1:4:6:4:1$). The three isozymes with intermediate mobilities observed in the heterozygotes are in fact hybrid molecules due to random association of polypeptide subunits encoded in the respective genes. The hybrid catalases can also be generated *in vitro*

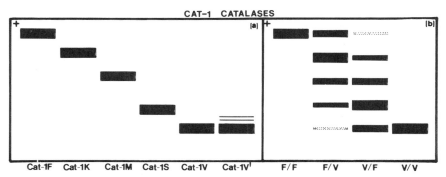

CAT-1 CATALASES

Fig. 4. (a) Schematic zymogram of the six CAT-1 catalase variants found in immature endosperm of different inbred strains of maize. Horizontal axis indicates alleles at the *Cat1* gene locus. (b) Shows two allelic variants (F and V) and the two kinds of heterozygote patterns obtained from reciprocal F_1 crosses (F × V and V × F). The apparent gene dosage effects are due to the triploid nature of the endosperm whereby there are two doses of the maternal gene and one dose of the paternal gene. The situation is similar with crosses involving any two alleles. Migration is anodal at pH 7.4.

by dissociation–reassociation experiments in $1 M$ NaCl (Scandalios, 1965). Thus, the genetic and biochemical data lead to the conclusion that the catalase of maize is structurally a tetrameric enzyme (Fig. 5).

C. Developmental Control of Gene Expression

Whether the same catalase phenotypes are expressed throughout maize development is an obvious question. Consequently an examination of the developmental time course and isozyme complement of catalase would provide the answers. Since the endosperm degenerates (utilized as a nutrient source) soon after germination, the scutellum was the tissue of choice for these initial experiments.

The developmental time course of scutellar catalase activity shows a significant increase in early sporophytic development with a peak around 3–4 days after germination (Fig. 6). Zymogram analysis further indicated that there are definite pattern shifts in catalase isozymes. In any given inbred strain homozygous for any allele at the *Cat1* gene locus, a series of "step-wise" shifts become apparent when scutellar extracts containing catalase are subjected to electrophoresis (insert, Fig. 6). The activity of the original CAT-1 isozyme, characteristic of the particular inbred strain during kernel development, decreases and gradually disappears, while

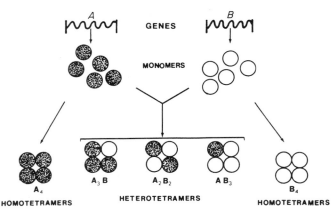

Fig. 5. Model for the formation of homotetramers and heterotetramers of catalase. Each gene (e.g., *A* and *B*) codes its respective monomeric polypeptide (subunit). If the monomers are allowed to aggregate randomly, five types of tetramers will be generated, two homotetramers (comprised of identical subunits) and three heterotetramers (comprised of mixed subunits). In diploid tissues the five species are distributed in a quantitative relation of 1:4:6:4:1 based on the binomial expansion of $(\frac{1}{2}A + \frac{1}{2}B)^4$. In a triploid tissue (i.e., endosperm) the distribution is skewed depending on the direction of the genetic cross $(\frac{2}{3}A\,♀ + \frac{1}{3}B\,♂)^4$.

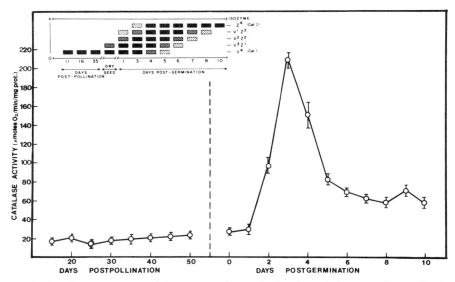

Fig. 6. Time course of catalase activity in developing kernels (days postpollination) and in scutella of germinating maize seeds (sporophytic development). The insert (top left) is a schematic zymogram showing the accompanying isozyme pattern shifts observed in scutella during the same developmental period. Note the gradual disappearance of CAT-1 (V⁴) and the gradual appearance of CAT-2 (Z⁴) with a period of overlap and generation of intergenic hybrid catalases.

that of the new isozyme CAT-2 increases and becomes the only form detectable 10 days after germination. The intermediate forms, observed at developmental stages when both the CAT-2 (Z⁴) and CAT-1 (V⁴) isozymes overlap in their expression, are not conformers of the CAT-1 isozyme; they are hybrid isozymes generated by the interaction of two kinds of subunits (i.e., V and Z polypeptides). This has been verified by *in vitro* hybridization of the two extreme (homotetrameric) forms, V⁴ and Z⁴ (Scandalios, 1975). Both the genetic and biochemical data clearly support the conclusion that the CAT-1 and CAT-2 isozymes are coded by two distinct and unlinked genes, *Cat1* and *Cat2* (Scandalios *et al.,* 1972). Genetic crosses between lines with electrophoretic variants of CAT-2 clearly show a Mendelian pattern of inheritance whereby the isozymes CAT-2Z and CAT-2P are coded by the respective allelic genes·*Cat2Z* and *Cat2P*. Thus in the maize catalase system, hybrid isozymes (heterotetramers) are generated by both intragenic (as in F₁ heterozygotes during kernel development) and intergenic (as in sporophytic development when both the *Cat1* and *Cat2* genes are expressed) complementation. This fact renders it an excellent system to study the possible physiological advantages or dis-

0

JOHN G. SCANDALIOS

advantages of heteromultimers over homomultimers (Scandalios *et al.*, 1972).

As mentioned above, the *Cat2* gene is initially expressed in the scutellum during early germination, but the timing varies somewhat with the particular inbred strain used, though not significantly (Scandalios *et al.*, 1972). Recently, however, we discovered that the *Cat2* gene is expressed as early as 25 days after pollination but only in the aleurone layer. Since both *Cat1* and *Cat2* genes are expressed simultaneously in this tissue, their subunit polypeptides interact to generate hybrid catalases. The activity is highest at 25 days and subsequently declines (Fig. 7). The aleurone is the outer layer of cells delineating the endosperm proper from the pericarp and is derived from the fertilization of the endosperm nucleus; thus it is triploid. The fact that it is normally comprised of a single thin layer of cells makes it a very difficult tissue to isolate and work with. However, a mutant has recently been derived (Wolf *et al.*, 1972) which has an aleurone comprised of five layers of cells instead of the one cell layer. This mutant should prove useful in our efforts to elucidate the

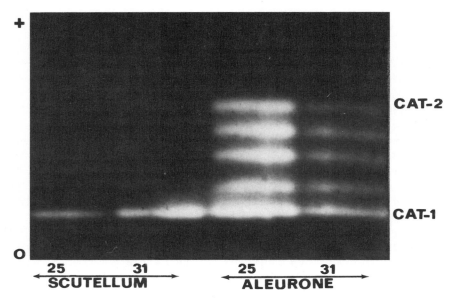

Fig. 7. Zymogram showing early expression of the *Cat2* gene in aleurone cells. Numbers on horizontal axis indicate day after pollination. Scutellum and aleurone were dissected from the same kernels at the respective days shown. The endosperm has only the CAT-1 isozyme expressed as in the scutellum. Note the generation of intergenic hybrid catalases in the aleurone where both the *Cat1* and *Cat2* genes are expressed simultaneously. 0 = point of sample insertion. Migration is anodal.

mechanisms controlling the early expression of the *Cat2* gene in the aleurone.

D. Regulation of Differential *Cat1* and *Cat2* Gene Expression

A number of posttranslational mechanisms have been found to affect the expression of the catalase genes in the scutellum during early sporophytic development. These mechanisms are described in some detail below.

1. Differential Turnover of the CAT-1 and CAT-2 Catalases

The differential expression of the *Cat1* and *Cat2* genes during development led us to the obvious question of whether the appearance of the CAT-2 and the simultaneous disappearance of the CAT-1 catalases were due to differences in turnover of the two isozymes. This question was answered by using the combined techniques of gel electrophoresis and density labeling in conjunction with isopycnic equilibrium sedimentation (Quail and Scandalios, 1971). Briefly, the method is based on incubating the tissue (i.e., scutellum in this case) with a protein precursor carrying a stable heavy isotope (^{15}N, ^{13}C, ^{18}O or $^{15}NO_3$, ^2H_2O, etc.) during the period that the enzymatic activity appears. The tissue is then homogenized and subjected to electrophoresis or electrofocusing; the buoyant density of the separated isozymes is then determined by isopycnic equilibrium sedimentation in a CsCl gradient. Purification of the enzyme is not essential. Mean densities can then be measured and compared to a marker enzyme. An increase in density of the enzyme is indicative of *de novo* synthesis during the period the tissue was incubated with the heavy isotope. Shifts from heavy isotope to light isotope (or vice versa) can be used to examine the simultaneous turnover of specific isozymes (Quail and Scandalios, 1971; Quail and Varner, 1971; Yang and Scandalios, 1975).

Our results clearly showed that both the CAT-1 (V^4 tetramer) and CAT-2 catalase (Z^4 tetramer) turnover in the scutella of maize seeds during germination; Z^4 as it accumulates, V^4 as it disappears (Fig. 8). We further demonstrated that the Z^4 tetramers accumulate because the rate of synthesis exceeds the rate of degradation, and V^4 tetramers disappear because the rate of degradation is in excess of their rate of synthesis. In addition, pulse-chase density labeling experiments indicate that the rates of synthesis of the CAT-1 and CAT-2 gene products are not equal; the Z^4 tetramer is synthesized at a greater rate than the V^4 tetramer (Quail and Scandalios, 1971). This may be a major factor in the differential expression of

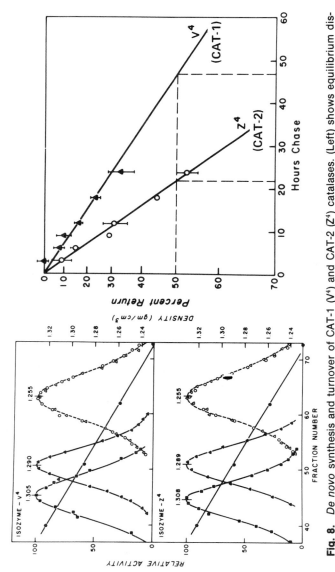

Fig. 8. *De novo* synthesis and turnover of CAT-1 (V⁴) and CAT-2 (Z⁴) catalases. (Left) shows equilibrium distribution in CsCl gradients of catalases V⁴ and Z⁴ from scutella grown for 36 hr on either $K^{14}NO_3$ in H_2O (▲——▲) or $K^{15}NO_3$ in 70% D_2O (■——■). The activity of the lactate dehydrogenase marker in the labeled (○——○) and unlabeled (△——△) gradients have been superimposed and drawn as one. Density of CsCl gradients (●——●). Relative activity means that all points on these curves are expressed as percentage of the highest point on each of the individual curves. (Right) shows time course of the decrease in the densities of the two isozymes during a 24-hr chase period after 36 hr of labeling. The fractional decrease in density is plotted as a percentage of the density difference ($\Delta\rho$) between the unlabeled and fully labeled molecules in each case and is designated as percent return. (With permission from Quail and Scandalios, 1971.)

these two catalase genes. Actual rates of synthesis and degradation were determined by an alternate method which involves following the decay of catalase activity after catalase synthesis has been blocked by the drug allylisopropylacetamide (AIA) (Price *et al.*, 1963). Since catalase activities in the scutellum are not constant during the experimental period in question, net synthesis could be determined directly by the increase or decrease in enzyme activity during a given 24-hr period. Degradation was determined by measuring the decrease in enzyme activity after blocking its synthesis with AIA. Total synthesis (k_S) was determined for each period by adding the net synthesis and the degradation. First-order degradation constants (k_D) were calculated by dividing the catalase activity (units) degraded during AIA treatment by the sum of the catalase activity at the beginning of the period and the total activity synthesized during the period. Half-lives of the enzymes were determined by dividing 0.693 (which is ln 2) by k_D (Rechcigl, 1971).

The rates of catalase synthesis increase to nearly three times the original value by the third day following germination. After day 4, the rates of catalase synthesis decrease to values similar to those of day 1 of germination (Table II). These data corroborate the density labeling results. Additionally, pulse-chase radiolabeling studies with [^{14}C]leucine (Sorenson *et al.*, 1977) did not indicate any changes in general scutellar protein turnover (Fig. 9), suggesting that the pattern of catalase degradation differs

TABLE II   Synthesis and Degradation of Maize Catalases during Early Germination[a]

Period of germination (days)	Control (no allyl-isopropyl-acetamide)	Net synthesis (units)	Degradation (units)	Synthesis (units)	k_D (units/ day/unit)
0	43.7 ± 7.3	—	—	—	—
1	90.7 ± 9.2	47.0	41.4 ± 4.4	88.4 ± 5.6	0.23
2	181.0 ± 7.7	90.3	39.3 ± 21.6	129.6 ± 11.5	0.13
3	363.9 ± 14.6	182.9	24.0 ± 3.3	206.9 ± 34.0	0.04
4	440.7 ± 19.3	76.8	127.2 ± 2.9	204.0 ± 34.2	0.20
5	410.5 ± 12.5	−30.2	146.5 ± 28.6	116.3 ± 12.1	0.28
6	371.2 ± 16.3	−39.3	118.6 ± 6.3	79.3 ± 14.3	0.26
7	330.3 ± 18.9	−40.9	126.1 ± 23.4	85.2 ± 13.2	0.30

[a] All values except k_D are expressed in catalase units per milligram fresh weight of scutellum. Degradation and synthesis values are listed as catalase units synthesized or degraded per milligram of tissue per day. Calculations are described in the experimental section. Standard deviations are based on a total of 7–12 replicates in three or four independent experiments. (From Sorenson *et al.*, 1977, with permission from the *Biochemical Journal*.)

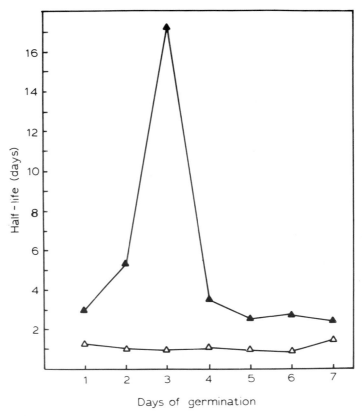

Fig. 9. Half-lives of catalase (▲) and general scutellar proteins (△) during early germination. Catalase half-lives were determined with the allylisopropylacetamide technique, and general protein half-lives were determined by pulse-chase labeling with [¹⁴C]leucine. (From Sorenson *et al.*, 1977, with permission from the *Biochemical Journal*.)

markedly from that of general scutellar proteins during the first 5 days of germination.

The rates of degradation observed during the developmental period examined change dramatically. Such changes may be due to alterations in the degradation mechanism or to variations in the susceptibility of the specific enzyme to degradation. Although it is difficult by present methods to distinguish between these two alternatives, we have some evidence for the latter explanation. The marked increase in catalase stability at day 3 may be due to the subcellular localization of the enzyme. A significant proportion of catalase (~50%) in the cell is found associated with glyoxysomes at days 3 and 4; these organelles are few in number before day 2, peak at day 4, and decrease sharply in number after day 4 (Scandalios, 1974; Longo and Longo, 1970). If we assume that the compartmentalized catalase is temporarily protected from degradation, the observed increase in apparent enzyme stability would be expected. After day 5, compartmentation of catalase is not significant, and after this point catalase degradation does not differ substantially from that of other scutellar proteins (Fig. 9), suggesting that catalase is probably not being degraded by a specific and unique system, although alternatives have not been eliminated unequivocally.

The turnover pattern for maize catalase appears thus to be consistent with the situation observed for most higher eukaryotes; it is a continuous process, and enzyme concentrations may be regulated by alterations in either the rate constants of synthesis or degradation (Schimke and Doyle, 1970).

2. Regulation by an Endogenous Catalase-Specific Inhibitor

a. Some Properties. The presence of a substance in maize which inhibits catalase activity has been reported (Scandalios, 1974). The inhibitor has been detected in most tissues of young seedlings, with highest activity being detected in the aleurone and the scutellum (Sorenson and Scandalios, 1976). The activity of the inhibitor varies inversely with catalase activity in the scutellum (Fig. 10) of the germinating seed, and constitutes one of several mechanisms regulating catalase activity in that tissue. We have purified the inhibitor by affinity chromatography on immobilized catalase. The purified inhibitor is trypsin sensitive and heat labile, indicating it is a protein. It has a molecular weight of 16,250 ± 1040 as determined by gel filtration on Sephadex G-100. The inhibition reaction proceeds fairly rapidly, being essentially complete within 5 min at 22°C. The inhibitor is not a protease (Sorenson and Scandalios, 1975).

b. Possible Mechanisms of Regulation. The inhibitor exhibits unusual kinetic properties (Fig. 11). This unusual pattern appears to be due to an enhancement of catalase activity at high substrate concentrations in the presence of inhibitor. The degree of enhancement is directly correlated with the amount of inhibitor in the assay mixture. One can envision three mechanisms (Sorenson and Scandalios, 1979) for such an enhancement.

1. It is conceivable that there could be some inacti ve catalase in the inhibitor preparation which is reactivated under high substrate conditions. This possibility is ruled out by the symmetrical protein elution profile of the purified inhibitor during gel filtration. Since the molecular

82

JOHN G. SCANDALIOS

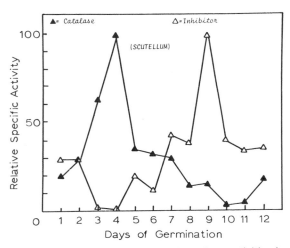

Fig. 10. Temporal expression of inhibitor and catalase activities in scutella of germinating maize seed. Each point is the mean of three independent experiments. Activities are given in units per milligram fresh weight and normalized with a peak value of 100.

weight of the catalase tetramer is 280,000 daltons (Scandalios *et al.*, 1972) and that of the subunit is 70,000 daltons, they differ substantially from the molecular weight of the inhibitor; thus contamination by catalase subunits (or any multiple thereof) should be readily detectable by gel filtration. Second, the inhibitor preparation shows no catalase activity, even when assayed under the high substrate conditions where the enhancement was observed.

  2. It is possible that the inhibitor could provide protection against enzyme inactivation by the substrate hydrogen peroxide. Such inactivation has, in fact, been well documented (Chance and Herbert, 1950). If the inhibitor acts by reducing the efficiency of a given enzyme molecule (as opposed to an all-or-none competitive or a dead-end type inhibition) and if the inhibitor also serves a protective function, one can envision a point where the protective effect exceeds the inhibitory effects. At this point, the activity of the enzyme plus inhibitor should be greater than the unprotected control. Higher levels of inhibitor should increase this effect. An analogous situation has been discussed by Segal (1975).

  3. It is conceivable that the inhibitor might exhibit a bifunctional effect such that it inhibits catalase action at low substrate concentrations, but enhances the enzyme activity at high substrate levels. Such a bifunctionalism could be due to an array of mechanisms (allosteric interactions between the inhibitor and hydrogen peroxide, for example); however, the

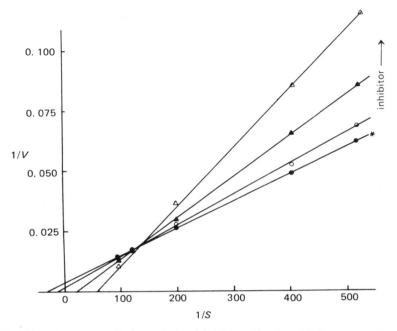

Fig. 11. Lineweaver–Burk analysis of inhibitor kinetics. All lines were fitted by least-squares regression analysis. The arrow indicates increasing concentration of inhibitor (39, 78, and 130 μg). Asterisk indicates control with no inhibitor added.

data obtained thus far do not warrant further speculation on the matter. This possibility is attractive in the biological context. Hydrogen peroxide is necessary for certain biological oxidations (Wolter and Gordon, 1975), and catalase would seem to be antagonistic to these processes under conditions of limited hydrogen peroxide. High levels of this substrate are deleterious, however, and under conditions of high peroxide production expeditious removal by catalase would be desirable (Fridovich, 1974). The existence of a molecule that could control this process thus seems attractive. We are currently examining these possibilities more closely.

c. Specificity. The inhibitor is effective against catalase from a wide variety of sources tested (Table III). These sources represent a wide evolutionary array, and the data would seem to indicate a degree of conservatism in catalase evolution. Immunological cross-reactivity between micrococcal and bovine liver catalases has also been demonstrated (Rupprecht and Schleifer, 1977), further establishing evolutionary conservation of enzyme structure. In light of the broad spectrum of sensitive catalases, it is not surprising that no difference is observed in the sensitivity of

TABLE III Activity of the Inhibitor against Catalase from a
Variety of Sources

Source	Sensitivity relative to maize[a]
Maize (*Cat1* locus)	100 ± 4
Maize (*Cat2* locus)	100 ± 2
Pea (*Pisum*)	102 ± 2
Algae (*Chlamydomonas*)	61 ± 5
Yeast (*Saccharomyces*)	101 ± 4
Bovine liver	100 ± 4
Drosophila	65 ± 3
Staphylococcus aureus	95 ± 8
Human erythrocyte	125 ± 3

[a] Maize catalase was inhibited 23% in these experiments.
Limits represent S.D. of three to five independent experiments.

the different maize isozymes to the inhibitor. However, the inhibitor is totally inactive against peroxidase or any other enzymes tested to date.

d. Inhibitor Variants. We have screened a substantial number of inbred maize lines for quantitative variation in levels of the catalase-specific inhibitor. Considerable variation was detected in levels of inhibitor activity, with the variation between extremes being approximately fifteenfold. Genetic crosses between low and high inhibitor level variants resulted in F_1 seed with low inhibitor levels behaving as a simple Mendelian dominant trait. These, however, are preliminary data.

It is becoming apparent that endogenous enzyme inhibitors may be an important aspect of the intracellular regulation of enzyme activity (Scandalios, 1977; Wallace, 1977). It is through further and more detailed studies of the catalase inhibitor that we may further our understanding of this regulatory strategy.

3. Intracellular Localization of CAT-1 and CAT-2

During kernel development, catalase is not associated with any distinct subcellular organelles. Whatever catalase is present is always associated with the soluble cytoplasm. However, as sporophytic development proceeds from the dry seed stage, glyoxysome biogenesis in scutella becomes apparent, reaching a peak at the fourth day of germination (Longo and Longo, 1970). Cell fractionation studies show that in scutella from young seedlings catalase is associated with the glyoxysomes and the cytosol (Fig. 12). It has been shown (Scandalios, 1974) that approximately 30% of the total catalase present in scutella at day 4 of germination is associated with the glyoxysomes; if we account for possible breakage of these organ-

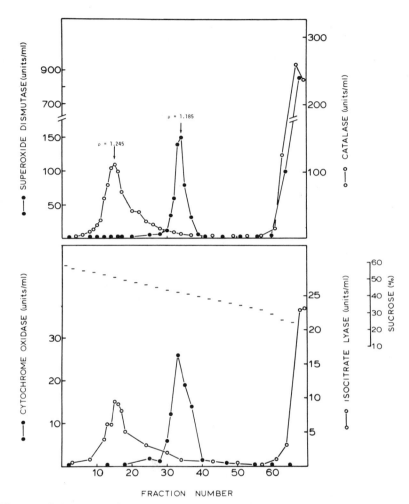

Fig. 12. Organelles from scutella of 4-day-old etiolated seedlings separated by sucrose gradient centrifugation. Isocitrate lyase and cytochrome oxidase are used as marker enzymes for glyoxysomes and mitochondria, respectively. Superoxide dismutase (SOD-3) is associated with mitochondria, while catalase is associated with glyoxysomes. Percentage sucrose is represented by the dashed line.

elles during isolation, the amount of associated catalase could be as high as 50% of the total, the remainder being cytosolic (W. F. Tong and J. G. Scandalios, unpublished). There is no detectable catalase associated with the mitochondria isolated from scutella at any developmental stage. There is no particular catalase isozyme specific to glyoxysomes (Scandalios, 1974); the isozyme complement of the organelles parallels that character- istic of the particular stage of scutellar development (Fig. 13; see also in- sert Fig. 6). Purity of fractions was determined enzymologically (using marker enzymes) and by electron microscopy (Fig. 14).

Recent iodination experiments using lactoperoxidase clearly indicate that the organelle-associated catalase is inside the glyoxysome structure and not merely bound to the outside surface of the organelles (P. H. Quail and J. G. Scandalios, unpublished). In addition, Longo *et al.* (1972), using the diaminobenzidine (DAB) histochemical reaction, demonstrated that the membranes of maize glyoxysomes become heavily stained after incuba-

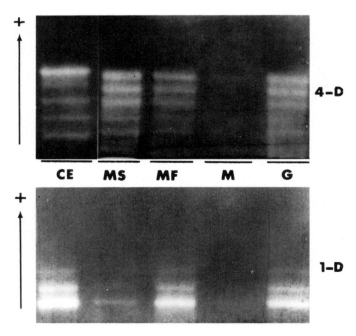

Fig. 13. Zymograms showing the distribution of catalase isozymes in different sub- cellular fractions isolated from maize scutella 1 day (1-D) and 4 days (4-D) after imbibi- tion. CE, crude extract; MS, mitochondrial supernatant; MF, mitochondrial fraction; M, mitochondria (after sucrose gradient centrifugation); G, glyoxysomes. Migration is an- odal. (From Scandalios, 1974, with permission from the *Journal of Heredity.*)

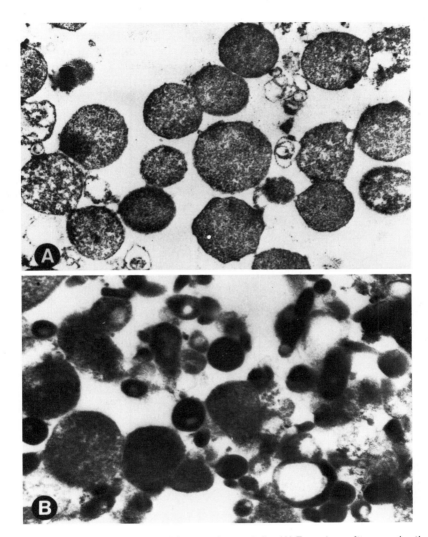

Fig. 14.   Glyoxysomes isolated from maize scutella. (A) Four days after germination. × 11,000. (B) One day after germination. × 24,000. Fixation in 3% glutaraldehyde and 1% osmium tetroxide; embedded in Epon. Note heterogeneity in 1-day fraction that consists mostly of electron-dense bodies. (From Scandalios, 1974, with permission from the *Journal of Heredity*.)

tion with DAB and H_2O_2. Their experiments also suggested that catalase may be present in the matrix as well as being membrane bound. The possible role of catalase in the function of glyoxysomes in converting fatty acids to sugars has been discussed elsewhere (Cooper and Beevers, 1969; Tolbert, 1971).

That the compartmentation process may have a stabilizing effect on catalase has been discussed (see Section IV,D,1).

E. Purification of CAT-1 and CAT-2 mRNAs

Although the CAT-1 and CAT-2 isozymes are differentially synthesized in the scutellum at the stated developmental period, there is no *a priori* reason to believe that the *Cat1* and *Cat2* genes are actively transcribing mRNA at the respective periods of catalase synthesis; the catalase synthesis observed could possibly reflect residual translation on a relatively stable mRNA.

In order to delineate the differences in the *Cat1* and *Cat2* genes responsible for their developmental regulation and synthetic capabilities, it is essential that comparisons be made of the relative rates of transcription, translation, and turnover of the mRNA species from these two genes. To this end, Dr. John C. Sorenson, my colleague in this department, has recently been examining catalase regulation at the translational level.

The specific mRNAs coding for the CAT-1 and CAT-2 catalase isozymes have been isolated from maize scutella. Monospecific antibodies have been used to specifically precipitate polyribosomes synthesizing either CAT-1 or CAT-2 catalase (J. C. Sorenson, personal communication). The mRNA was then purified from the precipitated polysomes by protease K digestion followed by oligo(dT) cellulose chromatography. Both of the messages are active templates for the synthesis of catalase protein in the wheat germ cell-free translation system. Electrophoretic profiles, using sodium dodecyl sulfate (SDS)-acrylamide gels, of cell-free translation products synthesized in response to the CAT-1 mRNA are shown in Fig. 15 (top). The single major polypeptide is judged to be catalase on the basis of its molecular weight and the fact that 85% of the ^{14}C incorporated into protein in this system can be precipitated with anti-CAT-1 antibodies.

The CAT-2 mRNA directs the synthesis of one major and several minor proteins (Fig. 15, bottom). The major protein has the same molecular weight as native catalase. Greater than 60% of the incorporated cpm can be precipitated with anti-CAT-2 antibodies. The lesser purity of the CAT-2 message is probably due to the fact that high ribonuclease content at the developmental stage used for CAT-2 mRNA isolation necessitates the use of suboptimal polysome precipitation conditions. Attempts to fur-

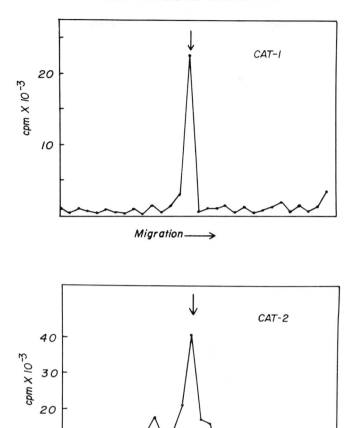

Fig. 15. Products synthesized in response to CAT-1 and CAT-2 mRNA preparations in the wheat germ cell-free translation system. Messages were purified by polysome immunoprecipitation followed by oligo(dT) cellulose chromatography. *In vitro* translation products were labeled with [^{14}C]leucine and electrophoresed in SDS–7.5% acrylamide gels. Following electrophoresis, gels were sliced into 2-mm slices and the radioactivity in each slice determined. Arrows indicate the migration of authentic CAT-1 and CAT-2 catalase purified from maize. (From J. C. Sorenson, unpublished data.)

ther purify this message are in progress. The purified messages will then be used to synthesize cDNA for use in studies of message transcription and turnover rates.

Thus, the catalase gene–enzyme system of maize appears to be amenable to studies at all levels of gene regulation. A particular advantage of this system in the scutellum is the fact that there is very little or no cell division occurring in this tissue. This allows for the continuous progression of cellular and subcellular physiology during development as opposed to an actively dividing system which periodically must shut down synthesis of normal metabolic proteins to concentrate on mitosis with the additional regulation a new spectrum of gene expression that process may involve.

F. Developmental Program Perturbations

The basic quantitative and qualitative profile of catalase activity during scutellar development (see Fig. 6) has been found to respond to external signals, both physical (i.e., light) and chemical (i.e., hormones). Additionally, there is now evidence that the activity levels of catalase during early sporophytic development are under genetic control.

1. Genetic Variants of the Developmental Program

We have recently recovered a variant maize strain with an altered catalase developmental program. When compared to strains exhibiting the standard catalase time course, the variant strain (397) exhibits a significantly different pattern with activity levels being higher in general and not declining after the fourth day of germination as is the case with the standard strains (Fig. 16). Both strains are identical with respect to the *Cat1* and *Cat2* structural genes. However, a higher rate of synthesis is maintained in strain 397 over a more extended developmental period as compared to W64A (D. Y. Chang and J. G. Scandalios, unpublished). The time course variation appears to be fairly specific, as it did not affect other enzymes examined. Furthermore, preliminary genetic analysis suggests that the difference between the two inbred strains is under genetic control (Fig. 17). The precise mechanism(s) by which this gene(s) regulates the expression and levels of the structural gene products (CAT-1 and CAT-2) is presently under rigorous examination in my laboratory.

2. Effects of Hormones on Catalase Gene Expression

Application of the plant hormone abscisic acid (ABA) to maize seedlings growing on nutrient media causes a significant change in the pattern of scutellar catalase activity during development (Fig. 18). In addition, zy-

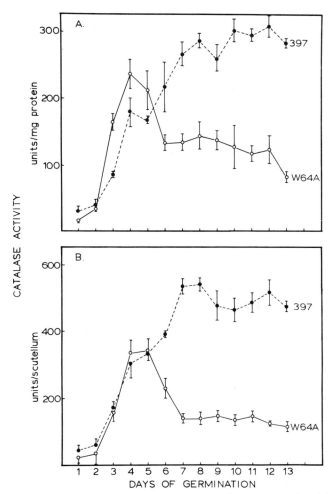

Fig. 16. Catalase developmental program variants. The pattern of activity from the inbred W64A is the one characteristic of all other inbred strains examined to date. The variant strain 397 exhibits a significantly different developmental time course. Scutella were isolated from each strain at the appropriate times indicated and assayed under standard conditions. The differences are significant whether activity is plotted as units per milligram protein (A) or units per scutellum (B).

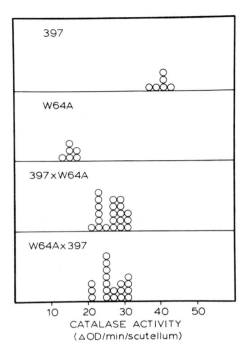

Fig. 17. Partial genetic analysis of catalase activity variants using isolated scutella from 10-day-old seedlings. Each circle represents one scutellum, and the position of each circle on the horizontal axis represents the catalase activity of each scutellum. The gene(s) controlling catalase activity is independent of the catalase structural genes, *Cat1* and *Cat2*, as both strains (W64A and 397) are homozygous for the same alleles at these loci.

mogram analysis showed that the expression of CAT-1, which normally disappears early in germination, is prolonged; while the initial expression of CAT-2 is somewhat delayed following ABA treatment. A number of other hormones tested (e.g., gibberellic acid, benzyladenine, and indole-acetic acid) do not appear to have a similar effect in altering the catalase developmental program (R. C. Yen and J. G. Scandalios, unpublished). The effect of ABA on the developmental pattern of catalase appears to be somewhat specific in that a number of other enzymes checked, i.e., peptidases, amylases, isocitratase, and several dehydrogenases, using the same extracts, do not respond similarly. The only exception was that the activity of alcohol dehydrogenase (ADH) also increases in response to ABA. Catalase activity in scutella excised from 6-hr soaked seeds increases threefold (over control) following ABA ($10^{-5}\,M$) treatment for 66-

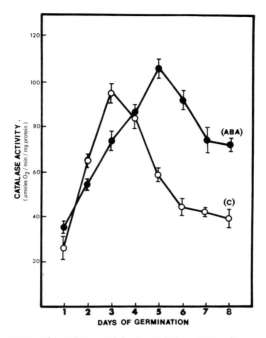

Fig. 18. Time course of catalase activity in maize scutella (C) and the effect of 10^{-5} M abscisic acid (ABA) in perturbing the catalase developmental program. S.D. is based on three replicates in four independent experiments.

hr in darkness. Density labeling experiments using ^2H_2O, indicate that the increase in catalase activity in response to ABA is due to *de novo* synthesis of the enzyme. The ABA treatment causes an increase in density shift by 0.0050 gm/cm^3 for catalase; these experiments also suggest that catalase may be turning over more rapidly than catalase in the controls.

The increase in catalase and alcohol dehydrogenase (ADH) activity in response to ABA may be physiologically significant. Abscisic acid has been reported to increase in response to anaerobiosis and may, in fact, mimic anaerobic conditions when applied exogenously (Beardsell and Cohen, 1975). It is further known that under anaerobic conditions the Krebs cycle is blocked, leading to accumulation of pyruvate. Excess pyruvate may then be decarboxylated to form acetaldehyde which, acting as a substrate, induces ADH (Scandalios, 1977). The ethanol generated in this reaction by ADH, may then be utilized as a donor substrate by catalase acting peroxidatically, to convert it to H_2O and acetaldehyde. The important aspect of this proposed cyclic scheme is that in the process

NAD^+ is regenerated from $NADH + H^+$, which is essential for further glycolysis and thus for germination. The scheme proposed is summarized in Fig. 19.

3. Effects of Light on Catalase Activity in Primary Leaves

In primary leaves of young seedlings catalase activity is relatively high during the initial 3–4 days after germination and subsequently declines (Fig. 20). On comparing light and dark grown seedlings, it was found that light promotes catalase activity after leaf development. The total activity of catalase in green (light grown) primary leaves increases very rapidly after 6 days of growth, while in etiolated (dark grown) primary leaves it remains low. Simultaneously, the specific and total activity of hydroxypyruvate reductase (a peroxisomal marker enzyme) increased very rapidly under light conditions. Other enzymes checked do not undergo similar changes.

To determine the effect of light on the subcellular distribution of catalase, the primary leaves of 5- or 6-day-old green and etiolated seedlings were used, and their homogenates were subjected to sucrose gradient centrifugation to fractionate the cells. As expected, catalase activity was found in the peroxisomal and cytosolic fractions. Additionally, catalase was found in association with the mitochondrial fraction. Thus, in the primary leaf, catalase is distributed among three subcellular compartments. Green primary leaves appear to have higher catalase activity in the peroxisomes than in the mitochondrial and cytosolic fractions, whereas in etiolated seedlings the reverse is the case (D. Y. Chang and J. G. Scandalios, unpublished).

These results suggest that peroxisome development in light is at least one factor which may be responsible for the elevated catalase activity in young maize leaves. This may be due to a stabilizing effect on the catalase molecules on binding to the microbody membranes, possibly protecting the enzyme from rapid degradation. It should be noted that the catalase

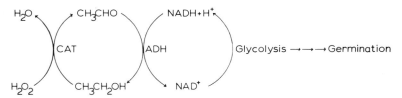

Fig. 19. Proposed cyclic reaction scheme in which catalase (CAT) and alcohol dehydrogenase (ADH) act cooperatively under anaerobic conditions to regenerate NAD^+ from $NADH + H^+$ to further glycolysis and stimulate seed germination (see text for discussion).

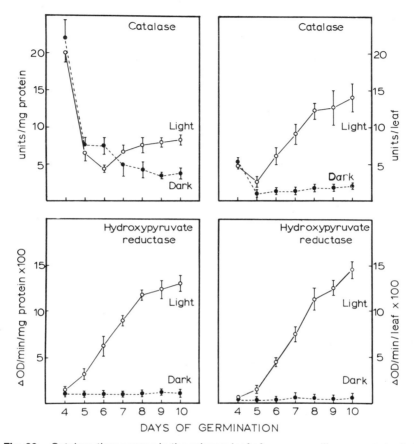

Fig. 20. Catalase time course in the primary leaf of young seedlings grown in darkness (etiolated) or light (green). Note the initial decrease in activity which is reversed under light conditions. Bottom shows effects of light on the peroxisomal marker enzyme hydroxypyruvate reductase. Activity is presented in both units per milligram protein and units per tissue.

present in the primary leaves (designated as CAT-3) proved to be uniquely different from the CAT-1 and CAT-2 isozymes; this is discussed in some detail in Section IV,G.

G. Tissue-Specific Catalases (Spatial Programming)

In addition to the temporal changes in catalase already discussed, there are a number of situations where a particular catalase appears in a given tissue (organ) at a given period in development. In the shoot of young

seedlings and in the pericarp of 25-day-old kernels, a unique catalase isozyme CAT-3 is found. Another isozyme CAT-4 is expressed in the pericarp on day 7 of sporophytic development (Scandalios, 1975). We have
recently, however, recovered an unusual strain of maize in which the
CAT-3 and CAT-4 isozymes are expressed in the pericarp at an earlier
stage of kernel development than is usually the case (Fig. 21). CAT-3 and
CAT-4 are expressed in the pericarp at the respective stage indicated
above, and the CAT-1 characteristic of the particular inbred strain is
usually, but not always, simultaneously expressed. When they are both
expressed, there are no apparent subunit interactions to generate hybrid
isozymes. Of the many inbred strains examined to date only two rare
cases have been detected in pericarp extracts in which CAT-1 and CAT-3
interact to generate hybrid isozymes (Scandalios, 1975). No such interactions have ever been observed between CAT-3 and CAT-2 or between
CAT-4 and any other of the known catalase forms. It should also be noted
that, with the exception of the unusual variant strain described above,
CAT-4 is very precise in its time of expression at the seventh day of germination (only in pericarp) with only minor traces of activity detected at

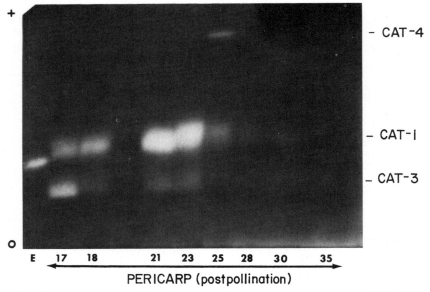

Fig. 21. Zymogram showing the precise time-dependent expression of the CAT-4
catalase. CAT-3 is usually expressed in the pericarp at the time indicated here and also
in young leaves. The expression of CAT-4 in this case is very rare; it is usually
expressed in pericarp at the seventh day after germination (see text, and Scandalios,
1975). E, milky endosperm.

the sixth and eighth days. Even in the variant strain shown in Fig. 21, CAT-4 is extremely precise in its time of expression (i.e., 25th day only). CAT-3 is normally the only catalase form expressed in the shoot of the young seedling. However, several lines have been found in which CAT-1 is expressed as a minor form compared to CAT-3. Similarly, we have recovered a few inbred strains in which CAT-3 is expressed as a very minor form compared to the predominant CAT-1 in the milky endosperm or in the scutellum of nearly mature kernels. Clearly, the genes controlling these forms, or the enzymes themselves, must be regulated differently in the different tissues; the precise control mechanisms involved are occupying much of our efforts. CAT-4, because of its almost fleeting existence is difficult to study in detail at this time; additionally, the pericarp poses some experimental problems. Consequently, we have focused some attention on CAT-3 from shoots of young seedlings for further study.

H. The CAT-3 Isozyme of Young Leaves

CAT-3 catalase has been studied in some detail using the epicotyl of young etiolated seedlings as the major source of material. This isozyme has a distinct electrophoretic mobility as compared to the CAT-2 and CAT-1 isozymes (Fig. 22). There is relatively low catalase activity present in the epicotyl of etiolated seedlings; less than 10% of that found in the scutellum.

1. Intracellular Localization

Following equilibrium centrifugation of epicotyl extracts on sucrose gradients; two peaks of organelle associated catalase activity were ob-

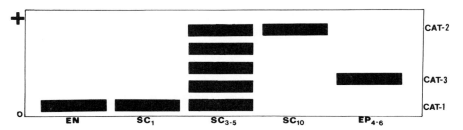

Fig. 22. Composite schematic zymogram of catalase in developing etiolated tissues of the inbred strain W64A. EN, milky endosperm at 16 days after pollination; SC_1, scutellum from 1-day imbibed seed; SC_{3-5}, scutellum from seeds at 3–5 days after germination; SC_{10}, scutellum at 10 days after germination; EP_{4-6}, epicotyls at 4–6 days after germination. CAT-1, CAT-2, and CAT-3 are the respective catalase phenotypes observed as indicated. 0, point of sample insertion.

served; one peak is associated with particles at a density of 1.21 kg/liter, which likely represents immature microbodies (i.e., peroxisomes). The second peak of catalase activity coincides with the mitochondrial peak at a density of 1.186 kg/liter and is tissue specific since it is found in epicotyls but not in scutella prepared from the same population of seedlings (Fig. 23). Mitochondria from the etiolated epicotyls were found to exhibit cyanide-insensitive respiration and, consequently, may require a catalase function (W. F. Tong and J. G. Scandalios, unpublished). Production of hydrogen peroxide by the alternate oxidase pathway (i.e., cyanide-insensitive respiration in mitochondria) has been recently reported for mung bean

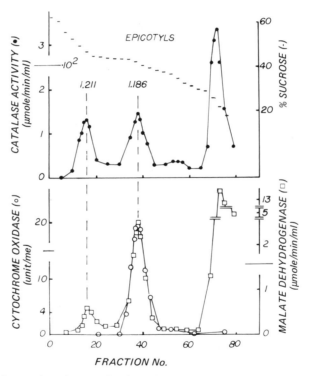

Fig. 23. Separation of organelles from 4-day-old etiolated epicotyls by stepped gradients (3 ml, 60%; 3 ml, 50%; 9 ml, 43%; 6 ml, 30% sucrose). One peak of catalase activity (ρ = 1.86 kg/liter) corresponds to the mitochondrial peak using cytochrome oxidase as marker enzyme and verified by purity of mitochondria by electron micrographs, and by the presence of only mitochondrial forms of malate dehydrogenase (MDH) isozymes in this fraction. A second fraction is associated with particles at a density of 1.21 kg/liter which probably represents immature peroxisomes since only peroxisomal MDH isozymes are associated with this fraction (see text).

hypocotyls (Rich *et al.*, 1976). Thus, coupling with the alternate oxidase pathway may be the primary role of catalase in mitochondria. Approximately 42% of the catalase activity present in the mitochondria was recovered in submitochondrial particles suggesting that the enzyme may be located in the intermembrane space of mitochondria. This result excludes the possibility that the mitochondrial-associated catalase could be due to microbody contamination. Furthermore, since the alternate oxidase has been reported to be located on the outer surface of mitochondrial inner membranes (Schonbaum *et al.*, 1971), it lends additional support that the CAT-3 catalase may be coupled with the alternate oxidase pathway.

2. Genetic Control of CAT-3

Two electrophoretic variants of CAT-3 have been found in shoots of a number of highly inbred lines screened. These variants have been labeled A and B. Epicotyls isolated from F_1 seedlings (A/B) show a five banded electrophoretic pattern indicating that CAT-3 is also a tetrameric enzyme. Detailed genetic analysis of the A and B variants suggests that they are the products of allelic genes at a single genetic locus *Cat3*. By use of A–B translocations and trisomic stocks we have located the *Cat1* gene on the short arm of chromosome 5 (D. G. Roupakias and J. G. Scandalios, unpublished). By similar analysis we are currently attempting to unequivocally identify the precise chromosomal location of the *Cat3* and *Cat2* genes.

I. Purification and Properties of Maize Catalases

The three isozymes of maize catalase CAT-1, CAT-2, and CAT-3 have been purified to homogeneity. All three catalases can be dissociated and reassociated *in vitro* by the method previously described (Scandalios, 1965) and all behave as tetramers. The polypeptides from the three genes have identical molecular weights (~280,000 daltons), but they differ significantly in a number of physicochemical or kinetic properties (Table IV).

In general we have found that CAT-2 and CAT-3 are more similar in their biochemical properties than either is to CAT-1. This may not be surprising if the *Cat2* and *Cat3* genes turn out to be located on the same chromosome and are likely to be more closely related evolutionarily. Alternatively, the *Cat1* gene is located on another chromosome and may have evolved more drastically in time. However, it should be noted that the CAT-1 and CAT-2 subunits interact, when both genes are operating in the same tissue at the same time (see Fig. 7), whereas CAT-3 subunits do not readily interact with CAT-1 or CAT-2 subunits. Whether this phenome-

TABLE IV Some Comparative Properties of the Purified CAT-1, CAT-2, and CAT-3 Isozymes of Maize[a]

Isozyme	Gene locus	No. of alleles[b]	Specific activity (units/mg protein)	K_m (H_2O_2)	Turnover rate[c] (hr)	Thermostability ($t_{1/2}$ at 50°C) (min)	Photosensitivity ($t_{1/2}$) (min)	AT inhibition[d] (% inhibition)	Subunit No.	Molecular weight (daltons)	pH optima	pI
CAT-1	Cat1	6(V)	561	0.143 M	47	7.0	6.8	82	4	~280,000	6–8	5.86
CAT-2	Cat2	3(Z)	2343	0.040 M	22	141.0	25.0	97	4	~280,000	6–8	5.01
CAT-3	Cat3	2(A)	688	0.062 M	22–24	12.6		99	4	~280,000	6–8	5.20

[a] All properties were determined under identical assay conditions for each pure enzyme.
[b] Letter in parenthesis indicates allele used to measure parameters in this table.
[c] Turnover rate, the time required to reach 50% saturation (50% of enzyme molecules are fully labeled) under our density labeling conditions.
[d] AT, aminotriazole (3-amino-1,2,4-triazole), a compound which irreversibly inactivates catalase by binding to the apoenzyme; 0.1 M at 37°C for 1 hr.

non has a genetic or a purely physiological basis is not yet clear; it is being investigated.

J. Gene Effects on Catalase Structure and Function: A Case for Molecular Heterosis

We have established that hybrid catalases can be generated by the interaction of allelic products, as in the case of heterozygotes between allelic genes at the *Cat1* locus, and by the interaction of nonallelic genes, as in the developmentally dependent expression of the *Cat1* and *Cat2* genes in scutella and aleurone (see Figs. 6 and 7). This situation provides an almost unique opportunity for the geneticist to study the effects of both intragenic (allelic) and intergenic (nonallelic) complementation on enzyme structure and function and to assess the possible physiological advantages and/or disadvantages of heteromultimers over the parental homomultimers. Such information is essential in any efforts to comprehend the molecular mechanisms underlying such concepts as single gene heterosis, fitness, codominance, etc. The biochemical complexity brought about by complementation may be selected in a natural population or at a given critical stage of an organism's development, since it may confer a definite advantage in an organism that is subjected to a wide range of environments as with a variety of biochemically differentiated tissues.

It has been demonstrated (Scandalios *et al.*, 1972) that the homotetrameric and heterotetrameric catalases of maize differ significantly with respect to their biochemical properties. In most instances the heterotetramers generated by either intragenic or intergenic complementation exhibit improved physicochemical properties over the least efficient parental molecules, suggesting that hybrid enzymes may be of advantage to the organism carrying them. Catalase activity in mixtures of two variants which were subjected to *in vitro* hybridization was higher as compared to simple (nonhybridized) mixtures (Fig. 24).

The data presented support the notion that the geometry of the subunits in catalase plays a significant role in the activity of the enzyme expressed. The nonallelic catalases CAT-1 and CAT-2 differ significantly in most properties measured, but they differ most dramatically with respect to two parameters: (1) the stability of the enzyme, as measured by the heat denaturation of catalase at 50°C; the CAT-1 allelic isozymes have a half-life of less than 10 min while the CAT-2 catalases have a $t_{\frac{1}{2}} > 140$ min, more than a 20-fold difference in thermostability (Fig. 25); and (2) the photosensitivity of the enzyme where CAT-2 was shown to be at least five times less sensitive ($t_{\frac{1}{2}} = 25$ min) to light than CAT-1 ($t_{\frac{1}{2}} = 6$ min). The half-lives of the hybrid isozymes reflect their subunit composition with in-

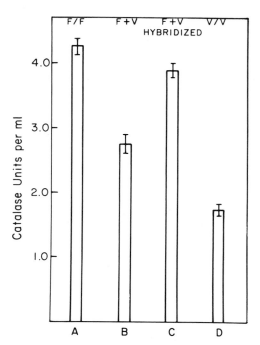

Fig. 24. Results of *in vitro* hybridization of the CAT-1 catalase homotetramers V^4 and F^4; extracts were obtained from the respective homozygous inbred strains. Catalase activity in the sample containing only $F^4 = 13.25 \pm 0.10$ units/ml; activity of $V^4 = 5.41 \pm 0.08$ units/ml. A mixture of 40% F^4 + 60% V^4 + 8.49 $\pm$ 0.08 units/ml. This compares well with the theoretical catalase activity expected of such a mixture: (0.4) (13.25) + (0.6) (5.41) = 8.54 units/ml. On subjecting the F^4 and V^4 extracts, as well as the mixture, to freeze-thawing in 1 *M* NaCl about 70% of the catalase activity is lost; where $F^4 = 4.24 \pm 0.09$ units/ml (A); V + 1.73 units/ml (D). If there were no interaction between F and V subunits, the expected catalase activity of the F^4 (40%): V^4 (60%) mixture would be (0.4) (4.24) + (0.6) (1.73) = 2.73 $\pm$ 0.09 units/ml (B). However, the actual catalase activity of this mixture is 3.89 $\pm$ 0.09 units/ml (C). (From Scandalios, 1975, with permission.)

termediate sensitivities (Fig. 26). In all instances, the CAT-2 catalases are comparatively more stable than the CAT-1 catalases, and the heterotetramers are catalytically more efficient than the least efficient parental homotetramer. These differences may be physiologically significant, since the *Cat2* gene is primarily expressed at germination and early sporophytic development when there is turning on of some major metabolic processes essential for development and differentiation of the maize plant; it is during this developmental period when the seed differentiates into the root and shoot.

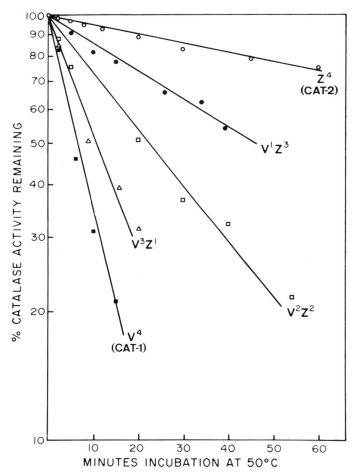

Fig. 25. Thermostability of the CAT-2 (Z^4) and CAT-1 (V^4) homotetramers and their interaction products (intergenic hybrid isozymes). Individual isozymes were partially purified, subjected to 50°C, and their activity measured at given time intervals. The percent catalase activity remaining at any given time was calculated by comparison to the original activity and the results plotted on log scale.

V. CONCLUDING REMARKS

The catalase gene–enzyme system described presents an attractive model for the study of posttranslational regulation of differential gene expression in a complex eukaryote. This system is well suited for studies on the genetic and epigenetic modulation of gene expression during de-

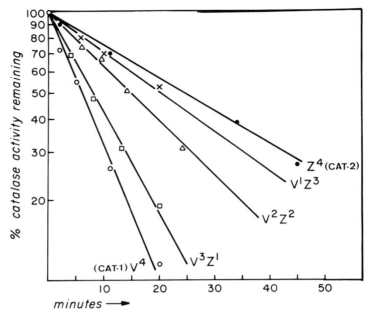

Fig. 26. Photosensitivity of the CAT-2 (Z^4) and CAT-1 (V^4) homotetramers and the hybrid isozymes generated by their subunit interactions. Each of the isozymes was partially purified and subjected to identical illumination conditions; activity was measured at definite time intervals. The percent activity remaining at any given time was calculated by comparing to the original activity and plotted on a log scale.

velopment since it meets the following criteria: (a) the enzyme exists in multiple molecular forms with characteristic physicochemical properties; (b) specific isozymes are detectable in single individuals and/or specific tissues; (c) the isozymes encoded in different genes exhibit significant temporal and spatial changes during maize development; (d) the different gene products are controlled by differing rates of synthesis and degradation, differential intracellular compartmentation, and by an endogenous catalase-specific inhibitor; (e) the system responds to and can be altered in its developmental expression by extracellular signals (e.g. light; hormones); (f) qualitative and quantitative mutations exist for catalase; and (g) maize is conducive to thorough genetic manipulation. The tissue specificity of catalase both qualitatively and quantitatively may be a reflection of the state of differentiation of a particular cell type or of its metabolic activity. Such "spatial programming" of gene expression also poses questions as to the mechanisms involved in developing and maintaining such differences by organisms whose somatic cells possess an identical genome (i.e., what mechanisms are involved in modulating gene activity in the diverse cell types in multicellular organisms?).

The fact that the *Cat1* and *Cat2* mRNAs have been successfully isolated, partially purified, and proved to serve as active templates for the *in vitro* catalase synthesis in a cell-free system clearly renders this gene–enzyme system amenable to regulation studies at all possible levels. The eventual purification and sequencing of the catalase genes themselves according to the general strategy recently applied in the purification of the ovalbumin gene (Anderson and Schimke, 1976) is certainly not inconceivable.

ACKNOWLEDGMENTS

I would like to express my sincere gratitude to my students and postdoctoral associates who participated in portions of the work reviewed in this chapter. I thank Mrs. Janet Barbour for typing the manuscript. I also thank a number of colleagues who critically read this manuscript. Research from the author's laboratory reviewed here has been supported by National Institutes of Health Grant No. GM 22733.

REFERENCES

Aebi, H., and Sutter, H. (1971). *Adv. Hum. Genet.* **2,** 143–199.
Anderson, J. N., and Schimke, R. T. (1976). *Cell* **7,** 331–338.
Beardsell, M. F., and Cohen, D. (1975). *Plant Physiol.* **56,** 207–212.
Beevers, H. (1969). *Ann. N. Y. Acad. Sci.* **168,** 313–324.
Bird, I. F. Cornelius, M. J., Keys, A. J., and Wittingham, C. P. (1972). *Phytochemistry* **11,** 1587–1594.
Chance, B. (1949). *J. Biol. Chem.* **179,** 1311–1330.
Chance, B., and Herbert, D. (1950). *Biochem. J.* **46,** 402–414.
Cooper, T. G., and Beevers, H. (1969). *J. Biol. Chem.* **244,** 3507–3513.
de Duve, C., and Baudhuin, P. (1966). *Physiol. Rev.* **46,** 323–357.
Eppenberger, H. M. (1975). *In* "The Biochemistry of Animal Development" (R. Weber, ed.), Vol. 3, pp. 217–255. Academic Press, New York.
Feinstein, R. N. (1970). *Biochem. Genet.* **4,** 135–155.
Freese, E. B., Gerson, J., Taber, H., Rhaese, H. J., and Freese, E. (1967). *Mutat. Res.* **4,** 517–531.
Fridovich, I. (1974). *Horiz. Biochem. Biophys.* **1,** 1–37.
Hendricks, S. B., and Taylorson, R. B. (1975). *Proc. Natl. Acad. Sci. U. S. A.* **68,** 1402–1406.
Koltzoff, N. K. (1927). *Zh. Eksp. Biol. Med.* **5,** 303–334.
Longo, C. P., and Longo, G. P. (1970). *Plant Physiol.* **45,** 249–254.

Longo, G. P., Dragonetti, C., and Longo, C. P. (1972). *Plant Physiol.* **50**, 463–468.

McCord, J. M. (1979). *In* "Isozymes: Current Topics in Biological and Medical Research" (M. C. Rattazzi, J. G. Scandalios, and G. S. Whitt, eds.), Vol. 3. Alan R. Liss, Inc., New York (in press).

Margoliash, E., Novgrodsky, A., and Schejter, A. (1960). *Biochem. J.* **74**, 339–350.

Markert, C. L. (1968). *Ann. N. Y. Acad. Sci.* **151**, 14–40.

Markert, C. L., ed. (1975). "Isozymes," Vols. 1–4. Academic Press, New York.

Markert, C. L. (1977). "Isozymes: Current Topics in Biological and Medical Research" (M. C. Rattazzi, J. G. Scandalios, and G. S. Whitt, eds.), Vol. 1, pp. 1–17. Alan R. Liss, Inc., New York.

Markert, C. L., and Møller, F. (1959). *Proc. Natl. Acad. Sci. U. S. A.* **45**, 753–763.

Nelson, M. S., and Scandalios, J. G. (1977). *J. Exp. Zool.* **199**, 257–268.

Nichols, P., and Schonbaum, G. R. (1963). *In* "The Enzymes" (P. D. Boyer, H. Lardy, and K. Myrbäck, eds.), 2nd ed., Vol. 8, pp. 147–225. Academic Press, New York.

Oshino, N., and Chance, B. (1973a). *Arch. Biochem. Biophys.* **154**, 117–131.

Oshino, N., and Chance, B. (1973b). *Arch. Biochem. Biophys.* **159**, 704–711.

Price, V. E., Starling, W. R., Tarantola, V. A., Hartloy, R. W., and Rechcigl, M. (1962). *J. Biol. Chem.* **237**, 3468–3475.

Quail, P. H., and Scandalios, J. G. (1971). *Proc. Natl. Acad. Sci. U. S. A.* **68**, 1402–1406.

Quail, P. H., and Varner, J. E. (1971). *Anal. Biochem.* **39**, 344–355.

Rechcigl, M. (1971). *In* "Enzyme Synthesis and Degradation in Mammalian Systems" (M. Rechcigl, ed.), pp. 263–310. Univ. Park Press, Baltimore, Maryland.

Rich, P. R., Boveris, A., Bonner, W. D., and Moore, A. L. (1976). *Biochem. Biophys. Res. Commun.* **71**, 695–703.

Rupprecht, M., and Schleifer, K. H. (1977). *Arch. Microbiol.* **114**, 61–66.

Scandalios, J. G. (1965). *Proc. Natl. Acad. Sci. U. S. A.* **53**, 1035–1040.

Scandalios, J. G. (1968). *Ann. N. Y. Acad. Sci.* **151**, 274–293.

Scandalios, J. G. (1969). *Biochem. Genet.* **3**, 37–79.

Scandalios, J. G. (1974). *Annu. Rev. Plant Physiol.* **25**, 225–258.

Scandalios, J. G. (1975). *In* "Isozymes: Developmental Biology" (C. L. Markert, ed.), Vol. 3, pp. 213–238. Academic Press, New York.

Scandalios, J. G. (1977). *In* "Regulation of Enzyme Synthesis and Activity in Higher Plants" (H. Smith, ed.), pp. 176–195. Academic Press, New York.

Scandalios, J. G., Liu, E., and Campeau, M. A. (1972). *Arch. Biochem. Biophys.* **153**, 695–705.

Schimke, R. T., and Doyle, D. (1970). *Annu. Rev. Biochem.* **39**, 929–973.

Schonbaum, G. R., and Chance, B. (1976). *In* "The Enzymes" (P. D. Boyer, ed.), 3rd ed., Vol. 13, pp. 363–408. Academic Press, New York.

Schonbaum, G. R., Bonner, W. D., Storey, B. T., and Bahr, J. T. (1971). *Plant Physiol.* **47**, 124–128.

Segal, I. H. (1975). "Enzyme Kinetics," pp. 182–192. Wiley, New York.

Shaw, C. R. (1969). *Int. Rev. Cytol.* **25**, 297–332.

Singer, T. P. (1971). *In* "Biochemical Evolution and the Origin of Life" (E. Schoffeniels, ed.), pp. 203–223. North-Holland Publ., Amsterdam.

Sorenson, J. C., and Scandalios, J. G. (1975). *Biochem. Biophys. Res. Commun.* **63**, 239–246.

Sorenson, J. C., and Scandalios, J. G. (1976). *Plant Physiol.* **57**, 351–352.

Sorenson, J. C., and Scandalios, J. G. (1979). *Plant Physiol.* (in press).

Sorenson, J. C., Ganapathy, P. S., and Scandalios, J. G. (1977). *Biochem. J.* **164**, 113–117.

Stern, C. (1968). *Jpn. J. Hum. Genet.* **13**, 181–182.

Takahara, S. (1952). *Lancet* **2,** 1101.

Takahara, S., and Miyamoto, H. (1948). *Jpn. J. Otol.* **51,** 163.

Tolbert, N. E. (1971). *Annu. Rev. Plant Physiol.* **22,** 45–75.

Wallace, W. (1977). *In* "Regulation of Enzyme Synthesis and Activity in Higher Plants" (H. Smith, ed.), pp. 176–195. Academic Press, New York.

Wolf, M. J., Cutler, H. C., Zuber, M. S., and Khoo, U. (1972). *Crop Sci.* 12, 440–442.

Wolter, K. E., and Gordon, J. C. (1975). *Physiol. Plant.* **33,** 219–223.

Yang, N. S., and Scandalios, J. G. (1975). *Biochim. Biophys. Acta* **384,** 293–306.

Yang, N. S., Sorenson, J. C., and Scandalios, J. G. (1977). *Proc. Natl. Acad. Sci. U. S. A.* **74,** 1310–1314.

3

Hormonal Control of Gene Expression

TUAN-HUA DAVID HO

PHYSIOLOGICAL GENETICS
Copyright © 1979 by Academic Press, Inc.
All rights of reproduction in any form reserved.
ISBN 0-12-620980-4

I. INTRODUCTION

In order to achieve a coordinated development, multicellular organisms employ the nervous system (only in animals) and hormones to integrate the function of their various organs. The term hormones can be generally defined as substances produced in minute quantities by one part of a multicellular organism, which are then transported to other parts of the organism to exert regulatory effects. In most of the vertebrates, hormones are produced in endocrine glands and carried by the circulatory system to various target tissues. Especially in mammals, there exists a very complex endocrine system in which the production of certain hormones is regulated by other hormones as well as by nerve impulses. A physiological process is often regulated by the integration of several different hormones. There are two types of animal hormones; peptides (such as insulin, glucagon, and growth hormones) and relatively small molecules (such as steroid hormones and thyroxine). In plants there are no specific endocrine glands, and some plant hormones are known to be transported through cells instead of being carried by body fluid. Essentially all the known plant hormones are relatively simple molecules, such as indoleacetic acid, ethylene, and gibberellins. However, it has been reported that a sexual inducer in the multicellular alga *Volvox* is a glycoprotein (Kochert, 1975). An unique feature of the plant hormone system is that sometimes a hormone [e.g., ethylene (Abeles, 1973)] can exert regulation in the cell where it is produced. In this case there is no distinction between the source and the target tissue of a hormone.

It is known that hormones can modify existing enzyme systems in the target tissue in order to increase their metabolic efficiency to meet the physiological demands. For example, when a mammal is under shock, a hormone epinephrine is produced by the adrenal medulla. After being transported by blood to its target tissue (liver), epinephrine binds to a specific membrane receptor of liver cells. As a consequence of this binding, a membrane-bound adenyl cyclase is activated and cAMP is produced. cAMP then activates a protein kinase which in turn activates existing glycogen phosphorylase. Glycogen stored in the liver cells is hydrolyzed by the activated glycogen phosphorylase. The resulting glucose is then released to the blood and distributed to muscle cells to serve as the energy source for a necessary response to the shock.

Hormones can also work as inducers of differentiation in target tissues. In this process, hormones derepress certain genes in the target tissues and induce the synthesis of new proteins in order to modify the function of target tissues. In this chapter I shall focus on the mechanism of this latter process. I shall review the current knowledge about the hormonal control of gene expression in the following five systems: induction of egg white proteins in the chick oviduct, vitellogenin synthesis in *Xenopus* liver, casein synthesis in the rat mammary gland, the synthesis of α-amylase in barley aleurone layers, and ecdysone in insect development.

II. HORMONAL INDUCTION OF EGG WHITE PROTEINS IN THE CHICK OVIDUCT

The chick oviduct is not only the passageway for egg migration but also the site where egg white proteins are synthesized and secreted. The mucosa of the magnum portion of chick oviduct is a tissue suitable for the study of hormonal control of gene expression. After exogenous estrogen stimulation, three distinct types of epithelial cells differentiate from the previously indistinguishable immature epithelial cells. In response to further hormone administration, the tubular gland cell produces ovalbumin, conalbumin, ovamucoid, and lysozyme, and the goblet cell synthesizes avidin. The third type of epithelial cell is ciliated and evidently concerned with the propulsion of material through the oviduct (Oka and Schimke, 1969; O'Malley *et al.*, 1969). If estrogen is withdrawn from chicks after the cytodifferentiation of oviduct has completed, the synthesis of egg white proteins is quickly resumed upon secondary stimulation with estrogen or progesterone (Fig. 1). The secondary hormone stimulation has been the subject of intensive molecular biological studies during the last decade. The sequence of events of hormone action in this tissue begins with the recognition of hormone molecules by the receptors. The hormone–receptor complexes are then transported to the nucleus where they bind to certain sites on the chromatin and subsequently stimulate the transcription of genes of egg white proteins (O'Malley and Means, 1974).

A. Specific Probes of Gene Activity

Ovalbumin is the most abundant protein synthesized (50–60% of total protein synthesized) in progesterone- or estrogen-treated oviducts (Palmiter and Schimke, 1973) and is often used as a marker to study the hormonal control of gene expression in this tissue. The activity of ovalbumin mRNA, as measured in a cell-free translation system, increases in the hormone-stimulated oviduct (Rhoads *et al.*, 1973; Chan *et al.*, 1973). Further, the specific mRNA of ovalbumin has been purified by conventional sizing methods with chromatography and electrophoresis (Woo *et al.*, 1975), or by a specific immunoprecipitation technique, in which the nascent peptide chains of ovalbumin synthesizing polysomes are first recog-

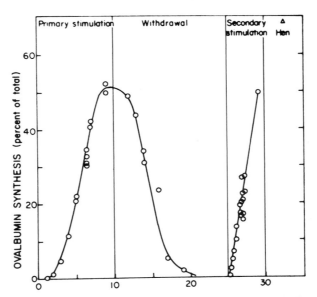

Fig. 1. Effect of estrogen on the relative rate of synthesis of ovalbumin in chick oviduct during primary stimulation, withdrawal, and secondary stimulation. (From Schimke *et al.*, 1975.)

nized and subsequently isolated by antiserum against ovalbumin (Palacios *et al.*, 1973; Shapiro *et al.*, 1974). The complementary DNA (cDNA$_{ov}$) of the purified ovalbumin mRNA has been synthesized by reverse transcriptase isolated from animal tumor viruses (Harris *et al.*, 1973; Sullivan *et al.*, 1973). cDNA$_{ov}$ can specifically hybridize with ovalbumin mRNA in a given RNA sample. From the hybridization kinetics and the specific radioactivity of cDNA$_{ov}$ one can calculate the number of sequences of ovalbumin mRNA in this particular sample. The ovalbumin mRNA and its precursors are the transcripts of the ovalbumin gene; thus, the number of sequences of ovalbumin mRNA with cDNA$_{ov}$ should reflect the activity of the ovalbumin gene.

The time course of ovalbumin synthesis in progesterone- or estrogen-treated oviduct correlates with that of the increase of the ovalbumin mRNA sequence as measured by cDNA$_{ov}$ hybridization (Fig. 2) (McKnight *et al.*, 1975). This observation together with other evidence (described later) seem to indicate that the increase of ovalbumin synthesis is mainly due to the expression of the ovalbumin gene. Palmiter (1975) has calculated that during the secondary estrogen stimulation the rate of synthesis of ovalbumin mRNA is between 22 to 34 molecules per minute per tubular gland cell. An obvious question is whether the ovalbumin gene is

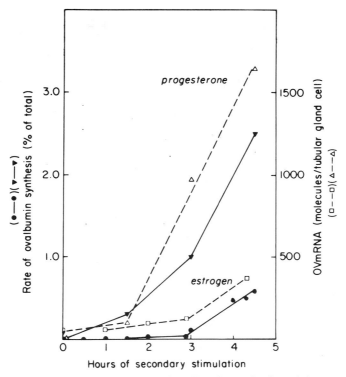

Fig. 2. Time course of the induction of ovalbumin synthesis and the accumulation of ovalbumin mRNA sequences during the secondary estrogen stimulation in chick oviduct. Ovalbumin mRNA sequences were measured by cDNA$_{ov}$ hybridization as described in the text. (From McKnight et al., 1975.)

amplified in hormone-stimulated chick oviduct in order to achieve such a fast rate of synthesis of ovalbumin mRNA. This question has been answered by employing cDNA$_{ov}$ to measure the number of ovalbumin genes in the genome of chick oviduct. The cDNA$_{ov}$ is allowed to hybridize with properly sheared oviduct DNA. The hybridization between cDNA$_{ov}$ and chick DNA has similar kinetics as the reannealing of unique sequences of chick DNA (Fig. 3) (Sullivan et al., 1973), indicating that the ovalbumin gene is not amplified in this tissue.

Thus far, only the activity of the ovalbumin gene has been discussed. How about the activities of other genes in the hormone-treated chick oviduct? In order to probe the overall activity of the whole genome, one has to first prepare cDNA against total cellular poly(A) RNA isolated from chick oviduct. The cDNA is allowed to hybridize with an excess of tem-

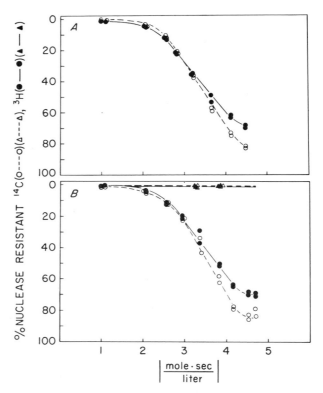

Fig. 3. Determination of the copy number of ovalbumin gene in the genome of chick oviduct. ^{3}H-Labeled cDNA$_{ov}$, ^{14}C-labeled unique sequence chicken liver, chick oviduct, or calf thymus were mixed together and then denatured and reannealed. At different times, aliquots were taken and assayed for the formation of double-stranded DNA (resistance to S$_1$ nuclease). The data are plotted relative to C$_0$t of the unlabeled cellular DNA. (A). ^{14}C-Labeled unique sequence DNA (○) and ^{3}H-cDNA$_{ov}$ (●) reassociated with chicken liver DNA. B. ^{14}C-labeled unique sequence DNA (○) and ^{3}H-cDNA$_{ov}$ (●) reassociated with chicken oviduct DNA; ^{14}C-labeled unique sequence DNA (△) and ^{3}H-cDNA$_{ov}$ (▲) reassociated with calf thymus DNA. (From Sullivan et al., 1973.)

plate poly (A) RNA. The kinetics of this hybridization depends on the sequence complexity of individual poly(A) RNA species. Using this approach, Monahan et al. (1976) have estimated that in estrogen-stimulated oviduct there are approximately 20,000 different poly(A) RNA sequences per cell, each representing only one to two copies per cell. However, there are five sequences which are present in a concentration of 5600 copies per cell. These five sequences are most probably the mRNAs of estrogen-induced egg white proteins. In the estrogen-withdrawn chick, there are only about 10,000 different sequences and the five abundant sequences are absent. Therefore, it seems that the withdrawal of estrogen not only stops the induction of egg white proteins, but also turns off many other genes. Hynes *et al.* (1977) used cDNA against *polysomal* poly(A) RNA to study the sequence complexity of poly(A) RNA being translated. In contrast to what was reported by Monahan *et al.* (1976), they found that there were only 13,000 mRNA sequences being translated in the estrogen-stimulated chick oviduct, and the withdrawal of hormone only caused the disappearance of abundant mRNAs without reducing the overall number of sequences. Taken together, the observations from these two groups of researchers suggest that estrogen induces not only the mRNA of egg white proteins, but also numerous other minor mRNAs. The frequency of these minor mRNA is low, and they are probably retained in the nucleus because they are not present on polysomes.

B. Hormone Receptors and Mechanisms of Gene Expression

The concept of steroid hormone receptors initially resulted from studies by Jensen and Jacobson (1962) in which physiological amounts of radioactive estradiol were injected into immature rats. It was noted by either biochemical or radioautographic methods that target tissue alone was capable of retaining radioactive hormone against a marked concentration gradient with blood. The chick oviduct, a specific target tissue for estrogen and progesterone, exemplifies a similar type of interaction with progestational steroids. When a chick treated with estrogen is subsequently injected with [^{3}H]progesterone, the major fraction of labeled steroid in oviduct cells is detected first in the cytoplasm and then the radioactivity is transported to the nucleus (Fig. 4) (O'Malley *et al.*, 1971). The radioactive steroid in the isolated hormone–receptor complex is identified as progesterone, not a metabolite of this hormone. The progesterone receptor can be extracted from cytoplasm; however, little or no salt-extractable progesterone-binding activity can be detected in the nuclei prior to progesterone administration. The receptor molecules appear to exist initially in oviduct cytoplasm. Following administration of [^{3}H]progesterone, a progressive increase in nuclear binding occurs (Fig. 5) (O'Malley *et al.*, 1971). The above observation indicates an absolute prerequisite for an initial interaction of the hormone with the cytoplasmic receptor and a concomitant depletion of cytoplasmic receptor as nuclear binding increases. The transport of hormone–receptor complex from cytoplasm to nucleus appears to be temperature dependent. The progesterone receptor protein has been highly purified (Schrader *et al.*, 1977). The hormone–receptor

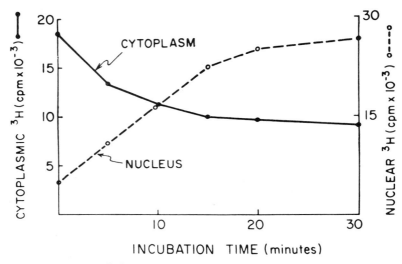

Fig. 4. Transfer of [³H]progesterone from cytoplasm to nucleus. Chick oviducts were incubated with [³H]progesterone for various periods as shown. The tissue was homogenated and cytoplasm and nuclei were separated. The radioactivity in these two cellular fractions were measured. (From O'Malley *et al.*, 1971.)

complex, either in crude preparation or in pure form, is capable of binding to isolated chromatin, thus, a great deal of information about the mechanism of gene expression by hormone–receptor complex is obtained by this type of *in vitro* study. The progesterone–receptor complex displays more extensive binding to oviduct chromatin than to the chromatin of chick spleen, heart, mature erythrocytes, or liver (Spelsberg *et al.*, 1971). Since the DNA of these different kinds of chick chromatin should be identical, the specificity of the binding probably has to reside on the proteins associated with chromatin.

There are two classes of chromatin proteins, histones and nonhistone proteins. Histones and most of the nonhistone proteins can be dissociated from DNA in the presence of high salt, and functional chromatin can be reconstituted by lowering the salt concentration. In this manner, "hybrid" chromatins are prepared with oviduct nonhistone proteins and histones from other tissues or species. Binding of hormone–receptor complex to this reconstituted oviduct chromatin with histones from nontarget tissue is similar to binding to the intact native chromatin of oviduct (Fig. 6) (Spelsberg *et al.*, 1971). Therefore, histones are not primarily responsible for the specificity of receptor binding. On the other hand, the tissue-specific binding can be transferred to nontarget tissue chromatin by exchanging the nonhistone chromatin proteins during reconstitution (Spels-

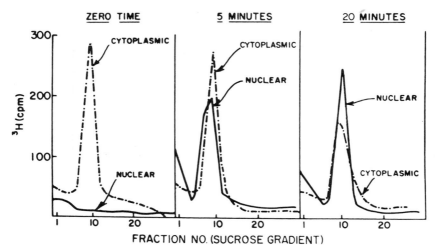

Fig. 5. Transfer of the [³H]progesterone–receptor complex from oviduct cytoplasm to nucleus. Tissues were incubated and homogenated as described in Fig. 4. Sucrose gradients were used to analyze the radioactivity profile, in the cytoplasm and the nuclear extract. The radioactivity peaks in the gradients represent [³H]progesterone associated with macromolecules (hormone–receptor complex). (From O'Malley *et al.*, 1971.)

berg *et al.*, 1971). A subfraction of the nonhistone proteins of oviduct chromatin (designated as P_3) has been found to be responsible for the binding of progesterone–receptor complex, because depletion of this protein fraction in chromatin causes the loss of the capacity to bind hormone–receptor complex (Spelsberg *et al.*, 1972). Progesterone receptor protein consists of two subunits (Schrader and O'Malley, 1972). The A subunit binds preferentially to naked DNA, and the B subunit binds only to chromatin (Schrader *et al.*, 1972). Together these subunits constitute a 6 S receptor dimer. It is postulated (Fig. 7) (Buller *et al.*, 1976) that following translocation to the nucleus, the receptor dimer binds through its B subunit to specific chromatin acceptor sites. Upon the binding of receptor dimer on chromatin, the A subunit is released. The released A subunit would then bind a nearby site. Because the A subunit binds to DNA directly, it may cause a destabilizing effect on the DNA duplex and expose more initiation sites for the RNA polymerase. This postulation is further supported by the following two observations. First, the rate of *in vitro* transcription of oviduct chromatin increases in the presence of progesterone–receptor complex (Schwartz *et al.*, 1976). Second, the number of initiation sites for RNA polymerase on isolated chromatin almost doubles within ½ hr after progesterone administration (Schwartz *et al.*, 1976).

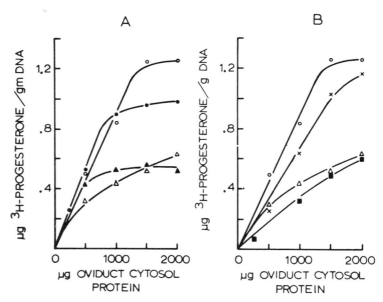

Fig. 6. Binding of progesterone–receptor complex to chromatin. Crude [³H]pro-gesterone–receptor complex from the cytoplasm of chick oviduct was used. (A) ●——●, Native chromatin from chick oviduct; ▲——▲, native spleen chromatin; ○——○, reconstituted oviduct chromatin with its own proteins; △——△, reconstituted spleen chromatin with its own proteins. (B) ○——○, Reconstituted oviduct chromatin with its own proteins; △——△, reconstituted spleen chromatin with its own proteins; χ——χ, reconstituted chromatin with oviduct DNA and nonhistone proteins and spleen histones, ■——■, reconstituted chromatin with spleen DNA and nonhis-tone proteins, and oviduct histones. (From Spelsberg *et al.*, 1971.)

Although the estrogen receptor in chick oviduct has not been inten-sively studied, a large amount of information about estrogen receptors has been obtained in other systems such as rat uterus. In the cytoplasm of rat uterus cells there are 4 S estrogen receptors. After the binding with the hormone molecule, the hormone–receptor complex is transported into the nucleus (Jensen and DeSombre, 1972). The nuclear form of estrogen–receptor complex has a sedimentation coefficient of 5 S, indicating a con-formational change of receptor protein may occur during the transport process. The 5 S nuclear complex, but not the 4 S cytoplasmic complex, can bind to chromatin (Jensen and DeSombre, 1972). Thus, the hormone-induced conformational change of the receptor molecule may be crucial to establish the binding capacity which is probably involved with the mecha-nisms of gene expression. Similar to the effect of progesterone, there is an overall increase in chromatin template activity in the estrogen-treated

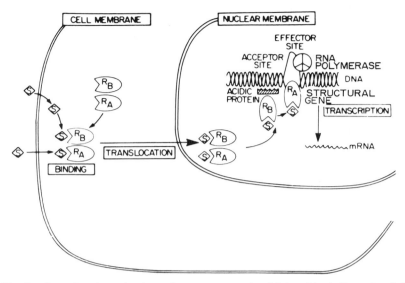

Fig. 7. Postulated mechanism of progesterone in chick oviduct. R_A, subunit A of the hormone receptor; R_B, subunit B of the hormone receptor. See the text for details. (From Buller *et al.*, 1976.)

chick oviduct. This results from the increased number of chromatin-binding sites for RNA polymerase and the generation of large numbers of new initiation sites for RNA synthesis on chromatin (Tsai *et al.*, 1975; Schwartz *et al.*, 1975). There is a close correlation between the concentration of nuclear estrogen receptors and the number of initiation sites for RNA synthesis, suggesting the possibility that gene transcription in this tissue may depend on the amount of estrogen receptor bound to the nuclei (Kalimi *et al.*, 1976). It is further demonstrated that the ovalbumin gene is among those genes whose initiation of transcription is enhanced by estrogen (Towle *et al.*, 1977). Garel and Axel (1976) have shown recently that more than 70% of the ovalbumin gene in chromatin isolated from hen oviduct can be digested by deoxyribonuclease, while regions in chromatin not actively transcribed are resistant to deoxyribonuclease. This result further indicates that the ovalbumin gene is somehow unmasked (probably uncovered by chromatin proteins) in estrogen-stimulated oviduct chromatin, so it is exposed to both RNA polymerase and deoxyribonuclease.

Nonhistone chromatin proteins also determine the specificity of estrogen-regulated transcription of the chick ovalbumin gene. Tsai *et al.* (1976) observed that nonhistone proteins from estrogen-stimulated chromatin are capable of activating the *in vitro* transcription of the ovalbumin gene when included in the reconstitution of chromatin from hormone-withdrawn oviduct. On the other hand, addition of extractable nonhistone proteins from withdrawn chromatin to stimulated chromatin does not affect the synthesis of ovalbumin mRNA.

C. Differential Induction of Different Egg White Proteins

Although ovalbumin and conalbumin mRNA accumulate in the same tubular gland cells of estrogen- or progesterone-stimulated oviduct, the kinetics of induction are different. The conalbumin mRNA starts to accumulate within 30 min of estrogen treatment, while the accumulation of ovalbumin mRNA has a lag period of 3 hr. Palmiter *et al.* (1976) have shown that the estrogen–receptor binding sites on chromatin are saturated within 15 min of hormone treatment. They suggested that a rate limiting translocation of hormone–receptor from the initial chromatin binding site to a more effective induction site, which governs the induction of individual mRNA, may determine the length of lag period of specific gene transcription. They further determined that the level of nuclear hormone–receptor complex for half-maximal induction of ovalbumin mRNA is about 2.5 times higher than the level required for half-maximal induction of conalbumin mRNA (Mulvihill and Palmiter, 1977). Therefore, there may be two types of induction sites on chromatin, one for ovalbumin and the other for conalbumin. The differential responses of ovalbumin and conalbumin induction may be related either to different number of specific binding sites regulating the induction of each mRNA, or to the different affinity of induction sites for estrogen receptors. More experimental evidence is needed to support this suggestion.

D. Future Experiments with the Ovalbumin "Minichromosome"

Although a great deal of information related to the induction of egg white proteins in oviduct has been gathered, the detailed mechanism of gene expression in this tissue will be revealed only by more delicate analyses in the future. The gene of ovalbumin, including the coding strand and the anticoding strand, has been isolated (Woo *et al.*, 1977a,b). With the help of recombinant DNA cloning techniques, a large quantity of this gene should be available. It is hoped that a "minichromosome" consisting of the ovalbumin gene together with nearby regulatory DNA sequences, histones, and nonhistone proteins can be constructed (O'Malley *et al.*, 1977).

In vitro interactions between the hormone–receptor complex and the "minichromosome" of ovalbumin should provide a more defined condition in which detailed mechanisms of gene expression can be studied.

III. VITELLOGENIN SYNTHESIS IN *XENOPUS* LIVER

Vitellogenin is the precursor of egg yolk proteins, phosvitin, and lipovitellin in egg-laying animals. This protein is normally synthesized in the liver and then transported through the blood to the ovary, where it is deposited in the developing oocyte. In the case of the South African toad, *Xenopus laevis*, vitellogenin has a molecular weight of approximately 200,-000 daltons. It contains phosphates and lipids and binds calcium ions. Vitellogenin is cleaved in the ovary to yield lipovitellin with a molecular weight of about 120,000 daltons, and phosvitin, an extensively phosphorylated protein with a molecular weight about 35,000 daltons in which 55% of the residues are serine (Clemens, 1974; Tata, 1976).

A. Induction of Vitellogenin mRNA

The biosynthesis of vitellogenin in liver is also regulated by estrogen. In female *Xenopus* the synthesis of vitellogenin is, of course, controlled by the level of endogenous estrogen. Male *Xenopus*, which do not normally produce vitellogenin, can also synthesize this protein after being injected with the hormone. Because the vitellogenin gene is not normally expressed in male *Xenopus*, it is a convenient system to study the effects of estrogen (exogenous) on gene expression. Attempts to obtain a full vitellogenic response to added estrogen in organ cultures of male *Xenopus* liver have been successful (Green and Tata, 1976; Wangh and Knowland, 1975). This further establishes that estrogen is the sole inducer of vitellogenin synthesis. However, so far most of the work has been done with intact male *Xenopus* injected with estrogen. The effect of estrogen on vitellogenin synthesis in liver is independent of hepatocyte cell division (Green and Tata, 1976). This is different from the effect of estrogen on ovalbumin synthesis in oviduct where administration of hormone initiates cell division and cytodifferentiation, resulting in the formation of new cell types capable of synthesizing ovalbumin and other specific proteins. DNA synthesis is not required for the initial stage of vitellogenin induction (Green and Tata, 1976). However, DNA synthesis may be necessary for the long-term maintenance of hormone-induced functions. The induction of vitellogenin synthesis by estrogen is fairly remarkable. The relative rate of vitellogenin synthesis is about 5% after 1 day of hormone treatment, and it reaches a maximum, which is more than 70%, at the twelfth day (Shapiro *et al.*, 1976). The level of translatable mRNA of vitellogenin as assayed in the reticulocyte lysate protein synthesis system essentially follows the same time course (Fig. 8) (Shapiro *et al.*, 1976). Due to the very large size of vitellogenin mRNA (30 S), this message is purified simply by size separation on sucrose density gradients (Wahli *et al.*, 1976; Shapiro and Baker, 1977). Baker and Shapiro (1977) used the cDNA synthesized against the purified vitellogenin mRNA as a probe to measure the number of vitellogenin mRNA sequence at different stages of estrogen treatment. They found that there was less than one molecule of vitellogenin mRNA in five liver cells during the first 3 hr of estrogen treatment. Afterward there was a fast and almost constant accumulation of vitellogenin mRNA, and there were about 36,000 molecules of this message per liver cell at the twelfth day of hormone treatment. This represents an impressive 180,000-fold increase of the amount of vitellogenin mRNA over a period of 12 days. If one calculates the rate of net synthesis of this mRNA, it is about 3000 molecules per day, or 120 molecules per hour. This rate is slightly slower than the rate of synthesis of ovalbumin mRNA in estrogen-treated chick oviduct. Since only one or two copies of ovalbu-

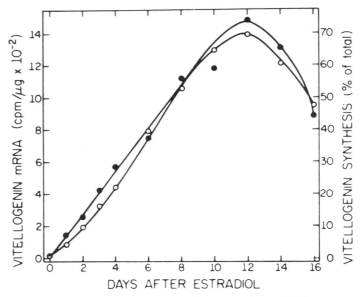

Fig. 8. Kinetics of estrogen induction of vitellogenin synthesis and vitellogenin mRNA. ○——○, Vitellogenin synthesis; ●——●, vitellogenin mRNA as measured by cell-free translation in reticulocyte lysate. (From Shapiro *et al.*, 1976.)

min genes are present in a haploid chick genome, it appears that the vitellogenin gene need not be amplified to accomplish this rate of mRNA synthesis. The observation that DNA synthesis is not required for the initial
induction of vitellogenin synthesis also tends to support this view. Ryffel
et al. (1977) using a similar approach have reported that the first appearance of vitellogenin occurs at about 12 hr after hormone treatment. The
difference between this report and the observation made by Baker and
Shapiro (1977) is probably due to the fact that Baker and Shapiro used
higher and more concentrated doses of estrogen. Higher concentration of
estrogen may saturate the estrogen receptors, thus making the transcription of vitellogenin gene the rate limiting step of vitellogenin induction.
The 3-hr lag period in vitellogenin gene transcription is roughly comparable to the 3-hr lag for the appearance of ovalbumin mRNA following secondary administration of estrogen to chick oviduct. Withdrawal of estrogen from *Xenopus* at the twelfth day causes a decline of the level of
vitellogenin mRNA (Baker and Shapiro, 1977). It decreases 12,000-fold
from day 12 to day 50 after the estrogen administration. At day 50 the
level of vitellogenin mRNA is about 3 molecules per cell and at day 60–65
the level of this mRNA is close to that of the unstimulated *Xenopus,* less
than 0.3 molecule per cell. As in the secondary stimulation of estrogen in
oviduct, readministration of estrogen to withdrawn *Xenopus* causes the
appearance of new vitellogenin mRNA at an initial rate approximately ten
times faster than that of primary hormone treatment. Baker and Shapiro
(1977) suggest that the primary estrogen treatment may result in some stable long-term changes in *Xenopus* liver cells, and this "memory effect"
causes either more rapid transcription of vitellogenin gene or an acceler-
ated transition to rapid transcription of the vitellogenin gene. This may be
due to the permanent increase of estrogen receptors or to a stable alteration of DNA or chromatin structure.

B. Deinduction of Serum Albumin

One interesting feature of estrogen-induced vitellogenesis in *Xenopus*
liver is the deinduction of other hepatic proteins. Liver cells normally
synthesize many proteins with serum albumin being the predominant one.
Green and Tata (1976), working with organ culture of male *Xenopus* liver,
found that the relative rate of albumin synthesis was more than 20% of
total proteins before the administration of estrogen. However, after hormone treatment the relative rate of albumin synthesis decreased dramatically (Fig. 9). The mechanism of this deinduction of albumin in estrogen-
treated liver cells is not understood. Is it due to the repression of the albumin gene or to the reduction of translation of existing albumin mRNA?

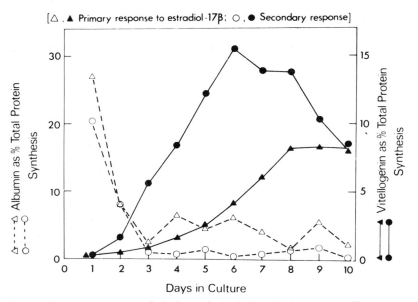

Fig. 9. Rate of synthesis of vitellogenin and albumin in cultured *Xenopus* liver. (From Tata, 1976, with permission of MIT press; copyright © MIT.)

Recently, rat albumin mRNA has been successfully purified by Schafritz and co-workers (Strair *et al.*, 1977). Similar techniques could presumably be employed to purify *Xenopus* liver albumin mRNA. With cDNA against albumin mRNA as a hybridization probe and the cell-free translation to assay albumin mRNA, one should be well equipped to study this intriguing problem. Information obtained from studies on the mechanism of this process should be valuable when compared with that related to the induction of vitellogenin. To date, most of the studies on hormonal control of gene expression have been concerned with the aspect of derepression of genes. Very few, if any, studies deal with the negative aspect of gene expression—how a hormone can turn off a gene. Thus, the deinduction of albumin synthesis in estrogen-treated *Xenopus* liver culture should offer a unique opportunity to gain some new insight into the regulation of gene expression.

C. Translational Control of Vitellogenin Synthesis

The estrogen-regulated vitellogenin synthesis in the rooster is somewhat different from that in male *Xenopus*. Estrogen injected into the rooster stimulates a fast synthesis of vitellogenin after a lag of 4 hr. The level of translatable mRNA of vitellogenin also increases in estrogen-treated rooster liver cells (Deeley *et al.*, 1977). After reaching its peak, the rate of vitellogenin synthesis declines, and there is little vitellogenin (2 to 5% of the peak) that can be detected in serum or liver cells 14 days after primary or 20 days after secondary hormone stimulation. However, there is still a significant amount of vitellogenin mRNA (about 30% of the peak value) present in liver cells at these stages (Mullinix *et al.*, 1976). This vitellogenin mRNA is still active in the reticulocyte lysate cell-free translation system and has essentially the same size as vitellogenin mRNA isolated from fully induced rooster liver cells (Mullinix *et al.*, 1976). This observation suggests that the 30% vitellogenin mRNA is not simply the residual activity of partially degraded mRNA. As expected, most of the remaining vitellogenin mRNA is not associated with polysomes but with 80 S monosomes (Mullinix *et al.*, 1976). Whether they are long-lived RNA masked in the form of ribonucleoprotein complex or they are under continuous turnover is still not clear. A more interesting unanswered question is whether these remaining vitellogenin messages in rooster liver cells are reused by the translation system to direct synthesis of vitellogenin following a subsequent stimulation of estrogen. Nonetheless, translational control, besides transcriptional control, of vitellogenin synthesis appears to be present in estrogen-stimulated rooster liver cells.

An observation supporting this notion was made by Beuving and Gruber (1970). They administered secondary estrogen stimulation together with actinomycin D to rooster at a time when low levels of vitellogenin were still present in the plasma. They found actinomycin D did not inhibit further accumulation of protein-bound phosphate (mostly vitellogenin-bound phosphate) in the plasma. Since vitellogenin is rich in serine (essentially in the portion of the molecule which becomes phosvitin), it is postulated that some new seryl-tRNA species could regulate the translational efficiency of vitellogenin mRNA (Tata, 1976).

IV. CASEIN SYNTHESIS IN RAT OR MOUSE MAMMARY GLAND

A. Cytodifferentiation of Mammary Gland

The growth and differentiation of the mammary gland is regulated by the interaction of multiple hormones. The mammary gland of rat or mouse prior to the onset of lactation consists largely of fat tissue with a branching ductal system of epithelial cells. The explants of mammary gland can be cultured in chemically defined medium (Turkington, 1972). The fat cells never proliferate in culture, while the epithelial cells can undergo cell divisions and differentiation in response to different hormonal treatment. Figure 10 diagrams the sequence of hormonal actions in epithelial cells of cultured mammary gland tissue. Insulin, epithelial growth factor, and growth hormones are capable of inducing the proliferation of epithelial cells (Turkington, 1972). The synthesis of DNA and RNA is enhanced by insulin (Turkington and Riddle, 1970), and the activity of RNA polymerase increases about threefold in insulin-treated cells (Chambon *et al.*, 1968). Histones and nonhistone proteins are phosphorylated (Turkington and Riddle, 1969), which is probably also related to the observed increase of transcriptional activity. Since it is known that insulin is able to enhance cAMP formation, protein phosphorylation may result from the activation of a protein kinase by cAMP. A suitable dosage of estrogen (10^{-12} to 10^{-10} *M*) promotes mammary gland development (Cowie, 1972; Turkington, 1972). However, at concentrations higher than 10^{-10} *M*, estrogen inhibits DNA synthesis in epithelial cells and prevents the proliferation of these cells (Turkington and Halif, 1968). Probably because of this inhibitory effect, estrogen is fairly useful in treating breast tumors which appear to result from the uncontrolled proliferation of mammary gland cells. Prolactin, a peptide hormone, is necessary to initiate casein synthesis in cultured mammary gland (Turkington *et al.*, 1967). Like insulin, prolactin enhances RNA synthesis and the phosphorylation of chromatin proteins (Turkington and Riddle, 1969). With the proper interactions among the hormones shown in Fig. 10, the epithelial cells can proliferate and become specialized in synthesizing and secreting milk proteins.

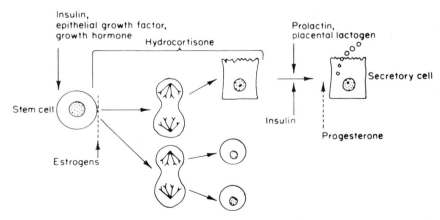

Fig. 10. Hormonal control of the differentiation of epithelial cells in cultured mammary gland. Solid arrows, promotion, dashed arrows, inhibition. (From Turkington, 1972.)

B. Induction of Casein

Caseins are the most abundant group of proteins (70–80% of total proteins) in milk. They are phosphorylated and rich in proline. There are three different caseins in rat mammary gland with molecular weights of 42,000, 20,000, and 25,000 daltons, respectively (Rosen et al., 1975). Using the wheat germ cell-free translation system, Rosen et al. (1975) have shown that the activity of the mRNA of all three caseins begins to increase before the fifth day of pregnancy and reaches a maximum during the lactation period (Table I). The activity of this mRNA declines after weaning when milk is no longer produced. To further quantitate the transcripts of casein genes, Rosen (1976) has purified the casein mRNA and prepared cDNA against the purified mRNA. With the cDNA as a hybridization probe, Rosen and Barker (1976) observed a coordinate increase in casein mRNA sequences and the mRNA activity as measured in cell-free translation system. This correlation favors the idea that the increase of the activity of casein mRNA during pregnancy and lactation period is due to the derepression of casein genes in mammary gland. The number of casein mRNA sequences on the eighth day of lactation has reached 79,000 molecules per alveolar (secretory) cell (Rosen and Barker, 1976). This value is within the same magnitude as the number of vitellogenin mRNA in *Xenopus* liver when its gene is fully induced by estrogen. The casein cDNA is further employed to show that, like the ovalbumin gene in chick oviduct, the casein genes are not massively amplified in mammary gland (Rosen and Barker, 1976).

TABLE I   Casein mRNA Activity during Mammary Gland Development[a]

Stage	Casein mRNA/ total mRNA[b] (%)	No. of molecules of casein mRNA/ aveolar cell[c]
5-Day pregnant	5.0	6,000
10-Day pregnant	11.4	37,000
13-Day pregnant	15.6	NA[d]
20-Day pregnant	25.3	55,000
2-Day lactating	40.3	NA
8-Day lactating	44.1	79,000
7-Day postweaning	1.7	NA

[a] Data taken from Rosen et al., 1975 and Rosen and Barker, 1976.
[b] Measured by *in vitro* translation.
[c] Measured by cDNA hybridization kinetics.
[d] NA, data not available.

C. Suppression of Casein Synthesis by Progesterone

Ovariectomy of the midpregnancy rat can increase the activity and number of sequences of casein mRNA by twofold (Rosen *et al.*, 1978). However, progesterone, but not estrogen or hydrocortisone, administered at the time of ovariectomy can prevent the increase in casein mRNA. Since progesterone is normally produced by the ovary, this observation indicates that the endogenous progesterone can prevent the full induction of casein genes during pregnancy. The synthesis and secretion of casein are not significant during midpregnancy, yet the level of casein mRNA is already about 25 to 50% of the maximal level reached during the lactation period (Rosen *et al.*, 1978). Does the endogenous progesterone in the mother also prevent the translation of casein mRNA during the pregnancy? The answer is probably yes because reducing the endogenous level of progesterone by ovariectomy can increase the synthesis of casein to about 40% of total proteins (Rosen *et al.*, 1978). Therefore, it appears that progesterone can prevent casein synthesis at both the transcriptional and the translational level during pregnancy. Since the progesterone level in mother drops substantially after the birth of baby (Cowie, 1972), this probably allows the full induction of casein genes and the efficient translation of casein mRNA and subsequently causes lactation.

V. HORMONAL REGULATION OF α-AMYLASE SYNTHESIS IN BARLEY ALEURONE CELLS

Barley aleurone consists of about three layers of nondividing cells surrounding the endosperm of barley seeds. During seed germination a hormone, gibberellin, is formed in the embryo and then transported to its target tissue, the aleurone cells. In response to gibberellin, the aleurone cells synthesize and secrete several hydrolases, including α-amylase and protease. After being secreted, these enzymes hydrolyze the reserved nutrients (starch and protein) in the endosperm. The resulting sugars and amino acids are then transported to the embryo to support the growth of the seedling (Yomo and Varner, 1971).

A. GA$_3$ Induction of α-Amylase

Isolated aleurone layers, which consist of only one type of cells, can respond to exogenous gibberellic acid (GA$_3$) (one of the gibberellins) in a fashion comparable to what happens during seed germination. In the presence of GA$_3$, the activity of α-amylase increases with a lag period of 4–8 hr (Fig. 11) (Chispeels and Varner, 1967). After 12 hr of hormone treatment, there is a linear and rapid accumulation of α-amylase indicating that the net formation of α-amylase has reached a constant and rapid rate (Varner and Ho, 1976). It has been demonstrated unequivocally that the increase of α-amylase activity is solely due to the *de novo* synthesis of enzyme molecules (Filner and Varner, 1967; Ho and Varner, 1978). α-Amylase has at least four different isozymes of identical molecular weight (Jacobsen *et al.*, 1970), and they together represent at least 45% of the newly synthesized proteins (Varner and Ho, 1976). The α-amylase gene(s) is probably not massively amplified during the hormone treatment because DNA synthesis inhibitors, such as fluorodeoxyuridine, cytosine arabinoside, and hydroxyurea, have no effect on the synthesis of α-amylase (Ho *et al.*, 1973). However, whether the four α-amylase isozymes are encoded from four different structural genes is still unknown. The synthesis of α-amylase can be inhibited by RNA synthesis inhibitors such as cordycepin if they are added at the same time as hormone administration (Ho and Varner, 1974). However, if the addition of cordycepin is delayed, its inhibition of α-amylase synthesis will decrease, and after 12 hr of GA_3

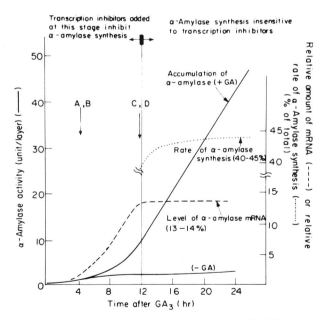

Fig. 11. Summary of GA_3 effects in barley aleurone cells. (A) Increase of poly(A) RNA synthesis. (B) Increase of the level of translatable α-amylase mRNA. (C) Secretion of α-amylase starts. (D) Cordycepin added at this time does not inhibit α-amylase synthesis.

treatment it does not inhibit the synthesis of α-amylase at all (Ho and Varner, 1974). Since most of the α-amylase is synthesized after 12 hr of GA$_3$ treatment, these results indicate that the mRNA of α-amylase is probably formed during the first 12 hr of hormone treatment. Ho and Varner (1974) observed that the synthesis of poly(A) RNA is enhanced by GA$_3$ during the first 12 hr. Higgins *et al.* (1976) translated barley aleurone mRNA in the wheat germ cell-free system and found that the level of translatable α-amylase. mRNA increased during the first 12 hr of hormone treatment (Fig. 11). The time course of these two events correlates with each other suggesting that the increase of α-amylase mRNA is likely due to the synthesis of this message.

B. Deinduction of Other Proteins by GA$_3$

Although the synthesis of α-amylase increases dramatically in the presence of GA$_3$, the rate of total protein synthesis remains more or less the same (Varner *et al.*, 1965). Analyzing the profile of newly synthesized proteins with SDS gel electrophoresis, D. Flint and J. E. Varner (personal communication, 1975) have observed that while the synthesis of α-amylase is increased, the synthesis of many other proteins is repressed by GA$_3$. This is further confirmed by the observation that the activity of an enzyme, arabinosyltransferase, decreases in the presence of GA$_3$ (Johnson and Chrispeels, 1973). Similarly, Ferrari and Varner (1969) have compared the substrate (NO$_3^-$) induction of nitrate reductase and the hormonal induction of α-amylase in barley aleurone cells. Nitrate reductase can be induced by its substrate, NO$_3^-$, in aleurone cells in the absence of GA$_3$. However, when GA$_3$ is present, the induction of nitrate reductase can no longer be observed. Therefore, GA$_3$ is able not only to turn on the synthesis of α-amylase but also to turn off the synthesis of other proteins. This is similar to the effect of estrogen on the shut-off of serum albumin synthesis in *Xenopus* liver cells as described in Section III.

C. Effect of ABA on α-Amylase Synthesis

Another interesting feature of hormonal regulation of α-amylase synthesis in barley aleurone layers is that all the GA$_3$ effects can be prevented by a second hormone, abscisic acid (ABA), which is able to suppress seed germination (maintain dormancy). Abscisic acid does not merely compete with GA$_3$, because a high concentration of GA$_3$ cannot overcome the inhibitory effect of ABA (Chrispeels and Varner, 1966). On the other hand, the effect of ABA appears to be very specific. If ABA is added 12 hr after GA$_3$ administration, when massive synthesis of α-amylase is already in progress, ABA can preferentially reduce the rate of synthesis of α-amylase (Fig. 12) (Ho and Varner, 1976). It has been suggested that ABA may regulate the translation of α-amylase mRNA (Ho and Varner, 1976); yet the possibility that ABA may also decrease the transcription of the α-amylase gene(s) has not been ruled out. In any case, the specific interaction of GA_3 and ABA on the synthesis of α-amylase deserves further investigation.

D. Genetic Approach

Mutants have been extensively used to study genetic and biochemical regulation in microorganisms but to a lesser extent in higher organisms. Apparently hormones have a complex action in barley aleurone cells as well as in other systems, and there are several regulatory steps involved. It would be beneficial if mutants blocked in one of the regulatory steps

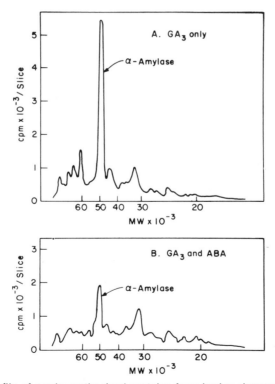

Fig. 12. Profile of newly synthesized proteins from barley aleurone cells. Proteins were analyzed by SDS gel electrophoresis. (From Ho and Varner, 1976.)

could be obtained for comparative studies. Such mutants could also be used to manipulate the network of genetic regulation of hormone action. Wheat aleurone cells, like barley aleurone cells, also respond to GA_3. A dwarf wheat variety (D 6899) whose aleurone layers produce very little α-amylase in the presence of exogenous GA_3 has already been found (Fick and Qualset, 1975). The partial insensitivity of this variety to GA_3 is regulated by a single locus (the *"Tom Thumb"* gene) on chromosome 4A. Since both the stem height (dwarfism) and the synthesis of α-amylase in aleurone cells are regulated by this gene (Fick and Qualset, 1975), it is possible that this gene regulates the level of hormone receptor or other crucial steps which are common to both stem elongation and the production of α-amylase. There are numerous existing wheat and barley varieties, and it would be beneficial to screen for more hormone-insensitive mutants among these varieties. It has been reported that a high frequency of mutation can be induced in barley by sodium azide (single locus mutation rate has been estimated to be $1/10^5$) (Kleinhofs *et al.*, 1978). Therefore, it is possible to increase the chance of scoring hormone-insensitive mutants by sodium azide mutagenesis.

VI. ECDYSONE IN INSECT DEVELOPMENT

A. Insect Hormones

Three major groups of insect hormones (ecdysones, juvenile hormones, and neurosecretory hormones) are known to regulate the morphogenesis of insects (Wyatt, 1972). Throughout the developmental process of insects, various genes are derepressed and repressed. However, there are very few cases in which a direct role of hormone on gene expression is observed. Among the many developmental events regulated by juvenile hormones are vitellogenesis and choriogenesis. The role of juvenile hormones on vitellogenin synthesis is similar to the effect of estrogen on the induction of vitellogenin synthesis in *Xenopus* or chick liver. The control of the synthesis of egg shell proteins during choriogenesis has been intensively studied by Kafatos and his co-workers (1977). No direct role of hormones has been found in the synthesis of egg shell proteins.

B. Ecdysone-Induced Chromosome Puffing

Ecdysone, commonly known as the molting hormone, is produced by the prothoracic gland of insects. In the developing insects, the prothoracic gland undergoes cycles of active hormone secretion and inactivity, which correlate with the timing of molting and metamorphosis, and in the adults the gland normally degenerated permanently. The function of this gland can be replaced by pure ecdysone.

It is known that ecdysone induces the sequence of puffing in the polytene chromosomes of the salivary gland of Diptera (e.g., *Drosophila*) (Clever, 1961; Berendes, 1967). Since the puffing and the regression of puffs in chromosomes have long been regarded as the activation and the inactivation of genes, it is apparent that ecdysone regulates gene expression during insect development. The puffing sequence at the late larval–early prepupal stage can be observed if the salivary gland is cultured in a defined medium supplemented with ecdysone, indicating that ecdysone is a primary inducer of the puffing activity. Withdrawing hormone from salivary gland, that has been exposed to ecdysone for 2 hr or longer, does not stop the puffing sequence (Ashburner and Richards, 1976). Therefore, the role of ecdysone seems to be the triggering of this sequence of events. What is the significance of chromosome puffing? Do the puffs eventually lead to the synthesis of hormone-induced enzymes (proteins)? An enzyme, dopa decarboxylase, rises sharply in activity following the increase in endogenous level of ecdysone (Shaaya and Sekeris, 1965). This enzyme is apparently involved in the phenolic tanning process of the puparium, which is regulated by endogenous ecdysone. The activity increase of this enzyme is accompanied by an ecdysone enhancement of RNA synthesis, but whether the specific mRNA of dopa decarboxylase is induced by ecdysone is not yet known. Analyzing the profile of proteins

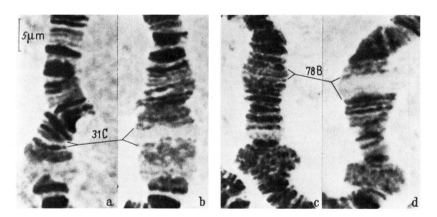

Fig. 13.   Ecdysone-induced chromosome puffing in *Drosophila*. (a) and (c) Chromosomes of control larvae. (b) and (d) Chromosomes of larvae treated with ecdysone. (From Berendes, 1967.)

synthesized in another tissue, the imaginal discs of *Drosophila*, by gel electrophoresis, Siegel and Fristrom (1974) observed that ecdysone treatment induced the appearance of several new protein bands. They found that the most dramatic effect of ecdysone was on the net synthesis of ribosomal proteins. Thus far, any correlation of the ecdysone-induced puffing at a particular locus on chromosomes and the appearance of a specific hormone regulated protein has not been reported.

Heat shock of cultured *Drosophila* cells can also induce chromosome puffing. Correlations of particular puff sites and the appearance of specific proteins and RNA have been observed (McKenzie *et al.*, 1975; Henikoff and Meselson, 1977). It is hoped that, in the future, a similar correlation can be obtained with the ecdysone-induced puffing.

VII. OTHER SYSTEMS

A. Induction of Metabolic Enzymes by Steroids

The induction of glutamine synthetase in neural retina (Moscona, 1973), the induction of tryptophan oxygenase in rat liver (Schutz *et al.*, 1973), and the induction of tyrosine aminotransferase in cultured hepatoma cells (Steinberg *et al.*, 1975) are all regulated by steroid hormones. The level of translatable mRNA of these enzymes is enhanced during hormone treatment. All of these enzymes represent less than 1% of total newly synthesized proteins; thus, the study of hormonal control of gene expression in these tissues is more difficult than those systems in which the marker proteins represent a major portion of newly synthesized proteins.

B. Mouse Lymphosarcomas

The mouse lymphoma line S49 is a system suitable for a genetic approach to hormone actions. The wild-type cells of this line are killed completely after a 48–72 hr exposure to physiological doses of glucocorticoids (Sibley and Tomkins, 1974). This allows an effective selection of mutants that are steroid resistant. So far, four resistant clones with varied properties of hormone receptor protein have been isolated (Yamamoto *et al.*, 1974; Yamamoto and Alberts, 1976). With the potential of having more mutant clones in the future, this cell line should be useful to pinpoint the genetic network that regulates a tissue's response to steroids.

C. Regulation of the Synthesis of Growth Hormone in Rat Pituitary Cells

Thyroid and glucocorticoid hormones stimulate the synthesis of growth hormone (a peptide) in cultured rat pituitary tumor cells. The synthesis of prolactin in these cells is not influenced by the presence of thyroid hormones (Seo *et al.*, 1977), indicating the effect of hormones is specific for the production of growth hormone. The level of growth hormone mRNA, as measured by both *in vitro* translation and cDNA hybridization, increases three- to fourfold after the administration of thyroid or glucocorticoid hormones (Martial *et al.*, 1977). This suggests that the gene coding for the growth hormone may be activated by thyroid and glucosorticoid hormones.

VII. SUMMARY

A scheme of the current knowledge about hormonal control of gene expression is shown in Fig. 14. It is a generally accepted view that any hormonal regulation, including gene expression, has to begin with the recognition of hormone molecules by the receptors in target tissues. The receptors can be located on the plasma membrane (e.g., insulin receptor), in the cytoplasm, or on the chromatin itself (e.g., ecdyson). In the case of estrogen and progesterone, the hormone molecules probably get into the

Target cell

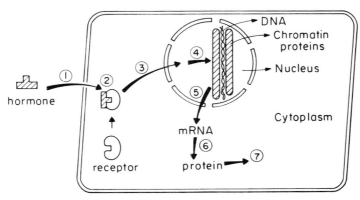

Fig. 14. Summary of the current knowledge of hormonal control of gene expression. See the text for details.

cytoplasm of target cells passively (step 1), where they are recognized by the receptors (step 2). The hormone–receptor complex is then transported to the cell nucleus (step 3). At the same time, some conformational change of the receptor proteins may take place, allowing the hormone–receptor complex to bind to the chromatin (step 4). The specificity of this binding is determined by certain fractions of nonhistone chromatin proteins. As a consequence of this binding, specific genes are expressed leading to the synthesis of specific mRNA (step 5). The occurrence of new mRNA will naturally change the spectrum of proteins synthesized (step 6) and subsequently modify the functions of target tissues (step 7). Posttranscriptional regulation may also take place to further amplify the effect of hormones on gene expression. For example, the ovalbumin mRNA and the *Xenopus* vitellogenin mRNA are very stable in the presence of hormone. However, these mRNAs quickly lose their stability when hormone is withdrawn. Although gene expression has been emphasized in this chapter, one has to keep in mind that the mode of action of hormones is not limited to transcriptional control.

ACKNOWLEDGMENT

This work was supported by a grant from The Jane Coffin Childs Fund for Medical Research (Project No. 61-468) and National Science Foundation Grant PCM 78-16143.

REFERENCES

Abeles, F. B. (1973). "Ethylene in Plant Biology." Academic Press, New York.
Ashburner, M., and Richards, G. (1976). *Dev. Biol.* **54**, 241.
Baker, H. J., and Shapiro, D. J. (1977). *J. Biol. Chem.* **252**, 8428.
Berendes, H. D. (1967). *Chromosoma* **22**, 274.
Beuving, G., and Gruber, M. (1970). *Biochim. Biophys. Acta* **232**, 529.
Buller, R. E., Schwartz, R. J., Schrader, W. T., and O'Malley, B. W. (1976). *J. Biol. Chem.* **251**, 5178.
Chambon, P., Karon, H., Ramuz, M., and Mandel, P. (1968). *Biochim. Biophys. Acta* **157**, 504.
Chan, L., Means, A. R., and O'Malley, B. W. (1973). *Proc. Natl. Acad. Sci. U. S. A.* **70**, 1870.
Chrispeels, M. J., and Varner, J. E. (1966). *Nature (London)* **212**, 1066.
Chrispeels, M. J., and Varner, J. E. (1967). *Plant Physiol.* **42**, 398.
Clemens, J. J. (1974). *Prog. Biophys. Mol. Biol.* **28**, 71.
Clever, U. (1961). *Chromosoma* **12**, 607.
Cowie, A. T. (1972). *In* "Hormones in Reproduction" (C. R. Austin and R. V. Short, eds.), p. 106. Cambridge Univ. Press, London and New York.
Deeley, R. G., Udell, D. S., Burns, A. T. H., Gordon, J. I., and Goldberger, R. F. (1977). *J. Biol. Chem.* **252**, 7913.

Ferrari, T. E., and Varner, J. E. (1969). *Plant Physiol.* **44**, 85.
Fick, G. N., and Qualset, C. O. (1975). *Proc. Natl. Acad. Sci. U. S. A.* **72**, 892.
Filner, P., and Varner, J. E., (1967). *Proc. Natl. Acad. Sci. U. S. A.* **58**, 1520.
Garel, A., and Axel, R. (1976). *Proc. Natl. Acad. Sci. U. S. A.* **73**, 3966.
Green, C. D., and Tata, J. R. (1976). *Cell* **7**, 131.
Harris, S. E., Means, A. R., Mitchell, W. M., and O'Malley, B. W. (1973). *Proc. Natl. Acad. Sci. U. S. A.* **70**, 3776.
Henikoff, S., and Meselson, M. (1977). *Cell* **12**, 441.
Higgins, T. J. V., Zwar, J. A., and Jacobsen, J. V. (1976). *Nature (London)* **260**, 166.
Ho, T. H. D., and Varner, J. E. (1974). *Proc. Natl. Acad. Sci. U. S. A.* **71**, 4783.
Ho, T. H. D., and Varner, J. E. (1976). *Plant Physiol.* **57**, 175.
Ho, T. H. D., and Varner, J. E. (1978). *Arch. Biochem. Biophys.* **187**, 441.
Ho, T. H. D., Keates, R. A. B., and Varner, J. E. (1973). *Plant Physiol.* **51**, s5.
Hynes, N. E., Groner, B., Sippel, A. E., Nguyen-Huu, M. C., and Schutz, G. (1977). *Cell* **11**, 923.
Jacobsen, J. V., Scandalios, J. G., and Varner, J. E. (1970). *Plant Physiol.* **45**, 367.
Jensen, E. V., and DeSombre, A. N. (1972). *Annu. Rev. Biochem.* **41**, 203.
Jensen, E. V., and Jacobson, H. (1962). *Recent Prog. Horm. Res.* **18**, 387.
Johson, K. D., and Chrispeels, M. J. (1973). *Planta* **111**, 353.
Kafatos, F. C., Regier, J. C., Mazur, G. D., Nadel, M. R., Blau, H. M., Petri, W. H., Wyman, A. R., Gelinas, R. E., Moore, P. C., Paul, M., Efstratiadis, A., Vournakis, J. N., Goldsmith, M. R., Hunsley, J. R., Baker, B. K., Nardi, J., and Koehler, M. (1977). *In* "Biochemical Differentiation in Insect Glands" (W. Beermann, ed.), p. 45. Springer-Verlag, Berlin and New York.
Kalimi, M., Tsai, S. Y., Tsai, M. J., Clark, J. H., and O'Malley, B. W. (1976). *J. Biol. Chem.* **251**, 516.
Kleinhofs, A., Warner, R. L., Muelbauer, F. J., and Nilan, R. A. (1978). *Mutat. Res.* **51**, 29.
Kochert, G. (1975). *In* "The Developmental Biology of Reproduction" (C. L. Markert and J. Papaconstantinou, eds.), p. 55. Academic Press, New York.
McKenzie, S. L., Henikoff, S., and Meselson, M. (1975). *Proc. Natl. Acad. Sci. U. S. A.* **72**, 1117.
McKnight, G. S., Pennequin, P., and Schimke, R. T. (1975). *J. Biol. Chem.* **250**, 8105.
Martial, J. A., Baxter, J. D., Goodman, H. M., and Seeburg, P. H. (1977). *Proc. Natl. Acad. Sci. U. S. A.* **74**, 1816.
Monahan, J. J., Harris, S. E., and O'Malley, B. W. (1976). *J. Biol. Chem.* **251**, 3738.
Moscona, A. A. (1973). *In* "Biochemistry of Cell Differentiation" (A. Monroy and R. Tsanev, eds.), p. 1. Academic Press, New York.
Mullinix, K. P., Wetekam, W., Deeley, R. G., Gordon, J. I., Meyers, M., Kent, K. A., and Goldberger, R. F. (1976). *Proc. Natl. Acad. Sci. U. S. A.* **73**, 1442.
Mulvihill, E. R., and Palmiter, R. D. (1977). *J. Biol. Chem.* **252**, 2060.
Oka, T., and Schimke, R. T. (1969). *J. Cell Biol.* **43**, 123.
O'Malley, B. W., and Means, A. R. (1974). *Science* **183**, 610.
O'Malley, B. W., McGuire, W. L., Kohler, P. O., and Korenman, S. G. (1969). *Recent Prog. Horm. Res.* **25**, 105.
O'Malley, B. W., Toft, D. O., and Sherman, M. R. (1971). *J. Biol. Chem.* **246**, 1117.
O'Malley, B. W., Towle, H. C., and Schwartz, R. J. (1977). *Annu. Rev. Genet.* **11**, 239.
Palacios, R., Sullivan, D., Summers, N. M., Kiely, M. L., and Schimke, R. T. (1973). *J. Biol. Chem.* **248**, 540.
Palmiter, R. D. (1975). *Cell* **4**, 189.
Palmiter, R. D., and Schimke, R. T. (1973). *J. Biol. Chem.* **248**, 1502.

Palmiter, R. D., Moore, P. B., and Mulvihill, E. R. (1976). *Cell* **8**, 557.
Rhoads, R. E., McKnight, G. S., and Schimke, R. T. (1973). *J. Biol. Chem.* **248**, 1870.
Rosen, J. M. (1976). *Biochemistry* **15**, 5263.
Rosen, J. M., and Barker, S. W. (1976). *Biochemistry* **15**, 5272.
Rosen, J. M., Woo, S. L. C., and Comstock, J. P. (1975). *Biochemistry* **14**, 2895.
Rosen, J. M., O'Neal, D. L., McHugh, J. E., and Comstock, J. P. (1978). *Biochemistry* **17**, 290.
Ryffel, G. U., Walter, W., and Weber, R. (1977). *Cell* **11**, 213.
Schimke, R. T., McKnight, G. S., and Shapiro, D. J. (1975). *Biochem. Actions Horm.* **3**, 245.
Schrader, W. T., and O'Malley, B. W. (1972). *J. Biol. Chem.* **247**, 51.
Schrader, W. T., Toft, D. O., and O'Malley, B. W. (1972). *J. Biol. Chem.* **247**, 2401.
Schrader, W. T., Kuhn, R. W., and O'Malley, B. W. (1977). *J. Biol. Chem.* **252**, 299.
Schutz, G., Beato, M., and Feigelson, P. (1973). *Proc. Natl. Acad. Sci. U. S. A.* **70**, 1218.
Schwartz, R. J., Tsai, M. J., Tsai, S. Y., and O'Malley, B. W. (1975). *J. Biol. Chem.* **250**, 5175.
Schwartz, R. J., Kuhn, R. W., Buller, R. E., Schrader, W. T., and O'Malley, B. W. (1976). *J. Biol. Chem.* **251**, 5166.
Seo, H., Vassart, G., Brocas, H., and Reeetoff, S. (1977). *Proc. Natl. Acad. Sci. U. S. A.* **74**, 2054.
Shaaya, E., and Sekeris, C. E. (1965). *Gen. Comp. Endocrinol.* **5**, 35.
Shapiro, D. J., and Baker, H. J. (1977). *J. Biol. Chem.* **252**, 5244.
Shapiro, D. J., Taylor, J. M., McKnight, G. S., Palacios, R., Gonzales, C., Kiely, M. L., and Schimke, R. T. (1974). *J. Biol. Chem.* **249**, 3665.
Shapiro, D. J., Baker, H. J., and Stitt, D. T. (1976). *J. Biol. Chem.* **251**, 3105.
Sibley, C. H., and Tomkins, G. M. (1974). *Cell* **2**, 213.
Siegel, J. G., and Friström, J. W. (1974). *Dev. Biol.* **41**, 314.
Spelsberg, T. C., Steggles, A. W., and O'Malley, B. W. (1971). *J. Biol. Chem.* **246**, 4188.
Spelsberg, T. C., Steggles, A. W., Chytil, F., and O'Malley, B. W. (1972). *J. Biol. Chem.* **247**, 1368.
Steinberg, R. A., Levinson, B. B., and Tomkins, G. M. (1975). *Cell* **5**, 29.
Strair, R. K., Yap, S. H., and Shafritz, D. A. (1977). *Proc. Natl. Acad. Sci. U. S. A.* **74**, 4346.
Sullivan, D., Palacios, R., Stavnezer, J., Taylor, J. M., Faras, A. J., Kiely, M. L., Summers, N. M., Bishop, J. M., and Schimke, R. T. (1973). *J. Biol. Chem.* **248**, 7530.
Tata, J. R. (1976). *Cell* **9**, 1.
Towle, H. C., Tsai, M. J., Tsai, S. Y., and O'Malley, B. W. (1977). *J. Biol. Chem.* **252**, 2396.
Tsai, M. J., Schwartz, R. J., Tsai, S. Y., and O'Malley, B. W. (1975). *J. Biol. Chem.* **250**, 5165.
Tsai, S. Y., Tsai, M. J., Harris, S. E., and O'Malley, B. W. (1976). *J. Biol. Chem.* **251**, 6475.
Turkington, R. W. (1972). *Biochem. Actions Horm.* **2**, 55.
Turkington, R. W., and Halif, R. (1968). *Science* **160**, 1457.
Turkington, R. W., and Riddle, M. (1969). *J. Biol. Chem.* **244**, 6040.
Turkington, R. W., and Riddle, M. (1970). *J. Biol. Chem.* **245**, 5145.
Turkington, R. W., Lockwood, D. H., and Topper, Y. J. (1967). *Biochim. Biophys. Acta* **148**, 475.
Varner, J. E., and Ho, T. H. D. (1976). *In* "The Molecular Biology of Hormone Action" (J. Papaconstantinou, ed.), p. 173. Academic Press, New York.
Varner, J. E., Chandra, G. R., and Chrispeels, M. J. (1965). *J. Cell. Comp. Physiol.* **66**, Suppl. 1, 55.

Wahli, W., Wyler, T., Weber, R., and Ryffel, G. U. (1976). *Eur. J. Biochem.* **66,** 457.
Wangh, L. J., and Knowland, J. (1975). *Proc. Natl. Acad. Sci. U. S. A.* **72,** 3172.
Woo, S. L. C., Rosen, J. M., Liarakos, C. D., Robberson, D. L., Choi, Y. C., Busch, H., Means, A. R., and O'Malley, B. W. (1975). *J. Biol. Chem.* **250,** 7027.
Woo, S. L. C., Chandra, T., Means, A. R., and O'Malley, B. W. (1977a). *Biochemistry* **16,** 5670.
Woo, S. L. C., Monahan, J. J., and O'Malley, B. W. (1977b). *J. Biol. Chem.* **252,** 5789.
Wyatt, G. R. (1972). *Biochem. Actions Horm.* **2,** 385.
Yamamoto, K. R., and Alberts, B. M. (1976). *Annu. Rev. Biochem.* **45,** 721.
Yamamoto, K. R., Stampfer, M. R., and Tomkins, G. M. (1974). *Proc. Natl. Acad. Sci. U. Sci. U. S. A.* **71,** 3901.
Yomo, H., and Varner, J. E. (1971). *Curr. Top. Dev. Biol.* **6,** 111.

4

Biochemical–Genetic Control of Morphogenesis

WILLIAM A. SCOTT

I. INTRODUCTION

As is well known, Beadle and Tatum used mutants to enormous advantage to elucidate biosynthetic pathways. Similarly, a biochemical–genetic approach has been extremely successful in delineating many regulatory mechanisms (Beckwith and Rossow, 1974), mechanisms of DNA replication (Kornberg, 1974), and mechanisms of self-assembly of the bacterio-

PHYSIOLOGICAL GENETICS

phage T_4 (Wood *et al.*, 1968) to name but a few examples. The analytical power of combining biochemistry and genetics to probe the molecular aspects of cellular processes cannot be overstated.

Numerous investigators have begun to examine the molecular events of differentiation by biochemical and genetic techniques. The underlying idea behind this approach is that the morphology of an organism is determined by its genetic constitution via enzymes and metabolic pathways. By analogy to biochemical mutants, therefore, morphogenetic changes can be studied by a comparison of mutants with the wild-type strain. For this and other reasons, prokaryotes and lower eukaryotes have attracted the attention of an increasing number of biologists and biochemists interested in problems of development and differentiation. The obvious advantage of these systems, in addition to the potential for genetic analysis within a short time span, include the relative simplicity of the developmental sequences compared to those of higher eukaryotes and an ease of handling.

Large numbers of mutants are a necessary prequisite for biochemical–genetic studies. Although selection procedures for development defective mutants in general are not available, efficient procedures can be devised for scoring large numbers of mutagenized cells for developmental defects. Arrested development or the inability to initiate development can usually be recognized by visual observation. Assuming differential gene expression is a universal feature of development such that different proteins are synthesized at specific stages of differentiation, three broad classes of mutations would be expected as found in *Dictyostelium* (Section IV,B). Mutations affecting both growth and development, mutations affecting only development, and mutations affecting only growth should be distinguishable. All three groups of mutants are of interest, since enzymes important only for the growth phase of the organism may suggest nonessential metabolic sequences for differentiation.

Formal genetic analysis can provide much information concerning the molecular events of development. Recombination studies permit a minimum estimate of the unique gene products to the differentiation phase of the life cycle. However, this type of analysis may yield inaccurate estimates of the relative number of gene products for the alternate life styles (growth and differentiation) of the organism unless a large sample size is employed and a random selection of mutants is assured. The spectrum of mutants obtained with different mutagens should indicate whether there is an equal probability for the occurrence of the three nonidentical mutant types with a frequency proportional to the actual number of genes. Analysis of such mutants can provide insight into the distribution and organization of the three classes of genes and, as a result, can suggest control mechanisms which regulate the expression of each.

In lower eukaryotes that exhibit distinctive stages of differentiation, it should be possible to correlate mutations with individual morphogenic events. The relative number of gene products unique to each developmental stage can be estimated by the techniques discussed above. Assuming the existence of unique gene products, mutants defective at one stage of development will complement with mutants affecting different differentiation events. The fact that this has been shown to occur (Sections II,A and IV,B) is indicative of "phase-specific" gene products. However, lack of complementation between mutants of sequential steps of the developmental pathway would suggest that a common gene product mediates both events. If these mutants carry out a normal developmental sequence in conjunction with mutants defective in other steps of development, then the gene product in question is critical to the original two steps only. Many possible sources of artifact can only be eliminated by a thorough examination of properties of the mutants. Nevertheless, from careful experiments of this type, the developmental sequence can be diagrammed as a map representing the expression of gene products mediating the various differentiation steps. Within the sequence, there may exist branch points where individual cèlls can follow alternative pathways. The molecular mechanisms whereby cells "decide" between pathways is of interest. Whether these decisions are in part or wholly genetically determined is unknown. Temporal mutants of *Dictyostelium* have been described (Sonneborn *et al.*, 1963) in which the time course of development is altered. This type of mutant suggests the existence of a genetic clock of the type proposed by Hartwell *et al.* (1974) as being operative in the cell cycle of yeast. Many other mutants with such interesting phenotypes are available, and more will become evident as the genetic analysis of development proceeds. These mutants are pertinent to the question of how an organism coordinates the events of development such that the sequence and timing remain invariant.

At a more biochemical level, mutational analysis of development can be utilized to ascertain which metabolic steps and pathways are critical to morphogenesis. The answer to this question requires a knowledge of the enzymatic lesions in developmentally deficient mutants. By elucidating enzymatic defects, the critical reactions to specific phases of development can be ascertained. Since phase-specific mutations are not corrected by supplementation techniques, it is unlikely that a nutritional requirement results in these phenotypes. Pathways most likely affected by these mutations, therefore, are those of intermediary metabolism and those concerned with the synthesis of effectors that initiate morphogenetic changes. Mutations resulting in an inability to transcribe phase-specific mRNAs must also be considered. A word of caution is warranted. Because the metabolism of a cell is an integrated entity, mutations of the first two types may have pleiotropic or multiple effects on the cell (Section II,D) and therefore influence unrelated areas of metabolism. As a result, the metabolic change responsible for the morphogenetic event may, in metabolic terms, be quite far removed from the primary enzymatic lesion (Scott and Mahoney, 1976). Although the developmental sequence of lower eukaryotes may be relatively simple, the biochemistry of these creatures can be formidable.

Obviously, the advantages and drawbacks of different organisms as model systems for studies of development depend on the questions to be asked and the experimental approach to be used. The three systems discussed in this chapter were chosen to illustrate this point and many of the points outlined above. The morphological mutants of *Neurospora* have been used as tools to probe the biochemical basis for morphogenetic change. *Mucor* is one of an interesting group of dimorphic fungi which can grow in either a yeastlike or hyphal state. Systematic genetic analysis has yet to be applied to this organism; however, a great deal is known about the factors which control dimorphism. As a result, mutants unable to effect the dimorphic change would greatly aid in elucidating the molecular basis of this relatively simple and reversible morphogenetic event. For this reason, *Mucor* is of interest to investigators with an inclination toward a biochemical–genetic approach to differentiation. The third system, *Dictyostelium,* has been a favorite model system for the study of development and has probably yielded more information concerning the molecular aspects of differentiation than any other organism.

II. *NEUROSPORA*

The life cycle of *Neurospora crassa* is well defined, and three phases can be distinguished: (1) vegetative growth, (2) asexual differentiation, and (3) sexual differentiation (Scott, 1976a; Schmidt and Brody, 1976). The morphological events associated with each phase have been described. In addition, the initiation of each phase can be controlled under defined culture conditions. Of the lower eukaryotic organisms utilized for studies of morphogenesis and differentiation, the advantage of *Neurospora* results from the vast amount of biochemical and genetic information available and the fact that this organism is amenable to analysis by both techniques.

A. Morphological Mutants

Much of the work with *Neurospora* related to differentiation has been concerned with the question of the biochemical basis of morphogenesis. Of the 500 or so known loci of *Neurospora*, mutations of at least 100 genes can lead to morphological abnormalities (Fig. 1). Bizarre patterns of vegetative growth result and are thought to occur because of increased hyphal branching in many of the so-called morphological mutants (Garnjobst and Tatum, 1967). The majority of the morphological mutations have pleiotropic effects on the life cycle of *Neurospora* in that the vegetative growth habit is affected in addition to asexual and/or sexual morphogenesis (Scott, 1976a). However, phase-specific mutants for conidiation or asex-

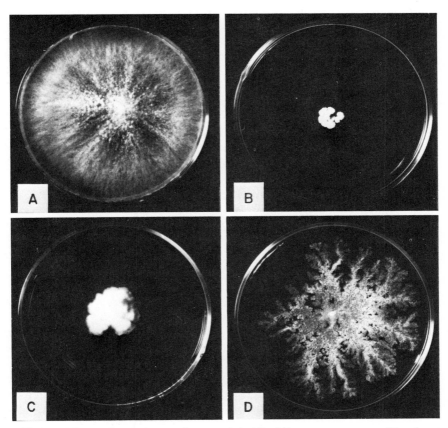

Fig. 1. Effects of mutation on the growth habit of *Neurospora crassa*. All cultures were grown for 2 days at 3°C. (A) Wild-type (B)–(D) Morphological mutants: *col-2, bal,* and *fr.* (Reproduced with permission from Scott and Tatum, 1979.)

ual development have been described (Selitrennikoff, et al., 1974). From linkage relationships, no evident clustering of morphological mutations is obvious except for those leading to a crisp phenotype (Garnjobst and Tatum, 1970). Loci for morphological mutants are known for all seven linkage groups of Neurospora (Garnjobst and Tatum, 1967).

Except for the temperature-sensitive mutants, environment has little influence on the phenotype of the Neurospora morphological mutants. These strains are prototrophic, further suggesting that mutations leading to morphological alterations do not affect catabolic pathways. The discovery several years ago that differences in cell wall composition between the morphological mutants and wild type (deTerra and Tatum, 1963; Mahadevan and Tatum, 1965) indicated that carbohydrate metabolism may be an important morphological determinant in Neurospora. This conclusion is based on the fact that the cell wall of Neurospora, like other fungi, is composed largely of polysaccharides together with glycopeptide components (Mahadevan and Tatum, 1965; Wrathall and Tatum, 1973). For changes in shape to occur, therefore, it is logical to assume that morphological alterations occur in part because of changes in cell wall structure. Based on this assumption, the primary enzymatic lesions in a number of morphological mutants have been traced to defective enzymes of carbohydrate metabolism (Section II,D).

The problem in elucidating the enzymatic reactions which control cell shape is one of identification. This difficulty is common to all systems employed for developmental studies. As implied above, knowledge of the composition of ultrastructural components and, in particular, differences in composition of components associated with phase-specific structures of the developmental sequence can greatly aid in this task. In Neurospora, numerous nutritional mutants are known, and the majority of these exhibit no morphological abnormalities under various supplementation conditions. As a result, the metabolic pathways affected by mutations leading to auxotrophy can be ruled out as being influential in morphogenesis. It should be emphasized that this does not imply that these reactions are not vital to the organism. In fact, most are essential for growth.

B. Choline and Inositol Auxotrophs

The exceptions to the rule that auxotrophic mutations have no effect on morphogenesis are most interesting. The choline (Beadle, 1944) and inositol (Shatkin and Tatum, 1961) auxotrophs behave as conditional morphologicals. Growth in media containing suboptimal levels of the growth factor (choline or inositol) results in morphological abnormalities. In contrast, the growth in fully supplemented media is phenotypically normal. In *Neurospora,* the choline-containing phospholipid, phosphatidyl-choline, accounts for approximately 50% of the total phospholipid of the cell. Minimal levels of the choline supplement lead to changes in the relative amounts of phospholipids in the choline mutant except for cardiolipin (Hubbard and Brody, 1975). This observation indicates that morphological changes of the choline mutant are accompanied by gross alterations in the lipid content of cellular membranes. Changes in membrane-associated functions are also known to occur when the choline requirement is not fully met by supplementation (Sherr, 1969; Lie and Nyc, 1962).

Supplementation of the inositol mutant with low levels of inositol also leads to changes in membrane phospholipid content. The level of phosphatidylinositol is reduced with a corresponding increase in the amount of phosphatidylserine (Hubbard and Brody, 1975). Levels of the other phospholipids are unaffected by varying the levels of the inositol supplement. Variations of these two minor phospholipids, phosphatidylinositol and phosphatidylserine, therefore, affect the growth habit of *Neurospora.* The morphological aberrations induced by low inositol may reflect only the initial events in the pathology of inositol deprivation which eventually leads to cell death (Lester and Gross, 1959).

Henry *et al.* (1977) have examined the events in yeast inositol mutants after starvation for inositol. On removal of the supplement, cells undergo division within 2 hr. However, no further mitotic division takes place, and macromolecular synthesis continues unabated without an increase in cell volume until death commences. It would appear that inositol deprivation results in an imbalance between the expansion of cell volume and the accumulation of cell constituents. Several observations support this idea. Density gradient centrifugation indicated that the density of inositol-minus cells was substantially greater than that of the supplemented population. Furthermore, cell death could be prevented by blocking cellular metabolism.

The essential role of inositol has yet to be satisfactorily explained. Since inositol occurs as a phospholipid component, it seems reasonable that the ubiquitous requirement for this sugar in eukaryotic systems (Mitchell, 1975) must result from the fact that membrane-associated activities function normally only if adequate concentrations of inositol phosphatides are present. That phosphatidylinositol is a necessary structural component, in the sense that it influences the physical properties of membranes, is an insufficient explanation for the importance of this phosphatide, since it is a minor constitutent of most membrane systems. A specialized role for phosphatidylinositol has been inferred from the observation that an enhanced turnover of the phosphorylinositol group of phosphatidylinositol is associated with the response of a variety of eukaryotic cells and tissues to stimuli (Mitchell, 1975). Similar increases in the metabolism of other membrane phospholipids are not observed under these conditions.

Several reports have indicated that phosphatidylinositol is obligatory to the normal functioning of membrane associated functions. For example, the $(Na^+ + K^+)$ATPase is severely impaired in inositol-deficient animals, and transport functions are dependent on an adquate supply of inositol (Mitchell, 1975). In addition to these activities, phosphatidylinositol is thought to mediate the stimulation of adenylate cyclase by norepinephrine (Levey, 1971a) and may also be required for normal levels of the basal activity (Levey, 1971b). Suboptimal levels of phosphatidylinositol, produced by supplementation of the inositol mutant of *Neurospora* with limiting levels of inositol, result in a reduction and destabilization of adenylate cyclase activity (W. A. Scott, C. Gabrielides, and J. Zrike, unpublished observations). Provided this decrease in activity reflects the *in vivo* situation, imbalance in the cyclic AMP economy of a cell may lead to a loss of metabolic regulation as found for the yeast mutant deprived of inositol. Whether a critical role for inositol phospholipids in membrane functions is a generalized phenomenon remains to be determined. However, Shapiro and Marchesi (1976) found that phosphorylation of human erythrocytes by inorganic [^{32}P]phosphate resulted in almost exclusive labeling of a polyphosphoinositide intimately associated with glycophorin A, the major protein of the red cell plasma membrane, suggesting this may be the case. In organisms such as those discussed here, the role of the plasma membrane in the synthesis and integration of cell wall components is an important consideration. The role of the cell surface in cell growth and morphogenesis is an area of which we unfortunately know little.

C. Cyclic AMP

Cyclic AMP is a central control element in determining growth and morphogenesis in *Neurospora*. This was indicated by two types of studies. In the first, phenocopies of morphological mutants were found to be produced by adding a variety of structurally unrelated drugs to wild-type *Neurospora* in liquid culture (Scott and Solomon, 1975). Drugs, such as atropine, histamine, theophylline, and several of the quinoline-containing antimalarials, including quinine, quinidine, and chloroquine) causes the wild type to assume a colonial or semicolonial growth habit. It should be noted that this is an unusual finding in that only a few compounds such as sorbose (Tatum *et al.*, 1949) and deoxycholate (Barber *et al.*, 1969) are known to induce morphological changes in *Neurospora*. One common effect of the former set of compounds is that all reduce intracellular cyclic AMP levels of the wild-type mycelia (Table I). Drugs that result in a colonial morphology lower cyclic AMP levels 70–75% compared to 40–50% for drugs that produce the less restricted spreading colonial morphology. Dose–response experiments also indicated a mutual relation between the degree of morphological restriction and the decrease in cyclic AMP levels. Reductions in cyclic AMP content could be correlated with the *in vitro* action of the drugs on the activities of the adenylate cyclase or the cyclic AMP-dependent phosphodiesterase. For example, quinine and atropine lower the substrate (ATP) binding constant of the adenylate cyclase. Presumably, cyclic AMP levels are reduced by decreasing the rate of cyclic AMP synthesis. On the other hand, histamine apparently increases the rate of cyclic AMP breakdown by activating (increase in V_{max}) the phosphodiesterase. One unusual finding was that theophylline, a well-known phosphodiesterase inhibitor (Robison *et al.*, 1971), inhibits the adenylate cyclase as well as the phosphodiesterase from mycelia grown in liquid cultures. Theophylline reduces endogenous cyclic AMP levels of mycelia by 40–5% suggesting that the predominant effect of the drug *in vivo* is the inhibition of the adenylate cyclase. This evidence indicates a relationship between the degree of morphological abnormality, the extent of the cyclic AMP reduction, and the effects of these drugs on the synthesis or breakdown of cyclic AMP. Although these drugs may have pleiotropic effects *in vivo,* the fact that similar effects on the cyclic AMP levels of wild-type *Neurospora* are observed on stimulation of the phosphodiesterase activity or on inhibition of adenylate cyclase nevertheless is striking.

With the cyclic AMP-deficient *fr* mutant, the question of whether a reduced cyclic AMP content results in morphological abnormalities can be approached by varying growth conditions and by supplementation tech-

TABLE I   Drug-Induced Changes in the Cyclic AMP Levels and Morphology of Wild-Type *Neurospora*[a]

Drug	Concentration (m*M*)	Cyclic AMP level (nmole/gm)	Morphology
None	—	4.2	Wild type
Quinidine	2.5	1.2	Colonial
Quinine	2.5	1.2	Colonial
Atropine	10	1.2	Colonial
Histamine	10	2.5	Semicolonial
Theophylline	10	2.5	Semicolonial

[a] Data from Scott and Solomon (1975).

niques (Scott, 1976b). By manipulation of the *fr* strain in this manner, a positive correlation between the cyclic AMP content and morphogenetic changes of *fr* can be shown, that is, progressive restoration of the morphological characteristics toward a wild-type state is paralleled by increases in cyclic AMP content. Table II indicates the cyclic AMP levels of *fr* and wild type under various growth conditions. Levels of the cyclic nucleotide increase from one-fifth in liquid cultures to one-half of the wild-type level in agar cultures and to normal levels on supplementation of agar cultures with phosphodiesterase inhibitors such as theophylline or ICI 63, 197. These changes in cyclic AMP content parallel the changes in the *fr* phenotype from a colonial form to a spreading colonial form and finally to a growth habit on theophylline–agar media that closely resembles the wild-type phenotype. The cyclic AMP level associated with a particular growth habit of *fr* is in excellent agreement with the comparable drug-induced morphology of wild type (cf. Tables I and II).

Theophylline and ICI 63, 197 increase the growth rate of agar cultures of *fr* threefold, but have little or no effect on the growth rate or growth habit of wild type. That these drugs should have beneficial effects on agar cultures of *fr* contrasts the results of the drug studies described above. Enzyme studies (Scott, 1976b) indicated that both drugs inhibit the phosphodiesterase from agar cultures of *fr* but have no effect on the adenylate cyclase. Similar results were obtained with the wild-type enzymes, indicating that this is not an unusual feature of the *fr* mutant. The dependence on growth conditions of adenylate cyclase sensitivity to purine analogues presents the interesting possibility that the state of this membrane-bound enzyme is different in mycelia grown in liquid and agar cultures.

Examination of the properties of the *fr* adenylate cyclase and phosphodiesterase suggested that a defect in the adenylate cyclase function may be responsible for the reduced cyclic AMP levels of *fr*. Striking differ-

TABLE II Cyclic AMP Levels of the *fr* Mutant and Wild-Type *Neurospora*[a]

Strain	Growth conditions	Cyclic AMP level (nmole/gm)	*fr* morphology
fr	Liquid	0.9	Colonial
Wild-type	Liquid	4.2	
fr	Agar	10	Semicolonial
Wild-type	Agar	20	
fr	Agar linolenate	20	Wild-type
Wild-type	Agar linolenate	20	

[a] Data from Scott (1976b).

ences in the sedimentation characteristics and the thermal stability of the mutant and wild-type adenylate cyclases are observed. Approximately 50% of the normal activity sediments at 105,000 g compared to 5% of the fr enzyme. In addition, the overall stability of the mutant adenylate cyclase is significantly decreased, and its rate of inactivation at 37°C in the absence of substrate is increased tenfold. Arrhenius plots also indicated that the Q_{10} and the temperature of maximal activity of the fr enzyme are reduced. The properties of the fr and wild-type adenylate cyclases from liquid and agar cultures were also compared to determine whether the properties of the fr enzyme parallel its functional capacity *in vivo* as indicated by the cyclic AMP levels of the mutant. It was found that the sedimentation behavior and temperature of maximal activity of the fr adenylate cyclase from agar cultures closely resemble those of wild-type enzyme, suggesting that there is a correlation between enzyme function and these properties. Nevertheless, differences between the fr and wild-type enzymes from agar cultures remain. In particular, the half-life of the fr adenylate cyclase is one-tenth that of the normal enzyme and is unaffected by culture conditions.

Supplementation experiments have also suggested an explanation for the adenylate cyclase defect of fr. Inclusion of polyunsaturated fatty acids in the growth medium of agar cultures results in an elevated cyclic AMP content and wild-type-like morphology (Table II) similar to that observed with phosphodiesterase inhibitors in addition to an adenylate cyclase with an increased thermostability (Scott, 1976b). Overall, the data are consistent with the idea that the reduction in cyclic AMP content of fr is caused by a defective adenylate cyclase and that the adenylate cyclase defect is a reflection of an altered membrane structure, presumably a lipid component. The fr mutant has a partial deficiency of linolenic acid (Section II,D) (Brody and Nyc, 1970), but whether this is the primary cause of the defective adenylate cyclase remains to be determined.

In addition to the fr strain, the crisp-1 (cr-1) mutant is also reported to have reduced adenylate cyclase activity and low cyclic AMP levels (Terenzi *et al.*, 1976). The cr mutants are characterized by abnormal mycelial growth patterns and premature conidiation (Garnjobst and Tatum, 1970). Exogenous dibutryl cyclic AMP and cyclic AMP are thought to promote the normal sequence of events leading to conidiogenesis in cr-1 since the cyclic nucleotide and its lipophilic derivative cause an increased production of aerial growth (Terenzi *et al.*, 1976). Some doubt exists as to whether this is the situation. If the result of a cyclic AMP reduction is to cause premature conidiation, the fr mutant and the drug-induced phenocopies of the morphological mutants should also have this phenotype. Since neither type of culture forms conidia unless growth is inhibited, it is

questionable whether early conidial production of *cr-1* is a consequence
of a cyclic AMP reduction. Other mutations (*cr-2* and *cr-3*) within the tight
gene cluster that leads to the crisp phenotype (Garnjobst and Tatum,
1970) do not have reduced adenylate cyclase activity (Torres *et al.*, 1971).
Furthermore, the activities of several enzymes are low in the crisp mutants (W. A. Scott and N. C. Mishra, unpublished observations). Because
of the frequent occurrence of modifier and morphological mutant genes in
vegetative cultures of crisp and in crisp strains recently isolated from
crosses (Garnjobst and Tatum, 1970) in contrast to other morphological
mutants, the finding of a defective adenylate cyclase in *cr-1* and its contribution to the phenotypic alterations of this strain are questionable.

D. Morphological Mutants with Defects of Carbohydrate Metabolism

The morphological mutants with known enzymatic defects are listed in
Table III. All of the enzymatic lesions described to date affect carbohydrate metabolism. This results from the bias of the investigators and
should not be construed to suggest that other metabolic sequences are un-
important to morphogenesis. Of the seven characterized lesions of morphological mutants, five are enzymes of the pentose monophosphate
shunt (PMS). Three unlinked mutations (*col-2, bal,* and *fr*) affect the glu-
cose-6-phosphate dehydrogenase, and two mutations (*col-3* and *col-10*)
affect the 6-phosphogluconate dehydrogenase (Table III). As discussed
previously (Scott and Mahoney, 1976), defects of the PMS should have

TABLE III   Morphological Mutants of *Neurospora* with Known Enzymatic Defects

Strain	Type of mutant	Enzyme defect[a]	References
col-2	Structural	G6P dehydrogenase	Brody and Tatum, 1966; Scott and Tatum, 1970
su-C	Suppressor	G6P dehydrogenase (?)	Scott and Brody, 1973
bal	Structural	G6P dehydrogenase	Scott and Tatum, 1970
su-B	Suppressor	None	Scott and Brody, 1973
fr	Structural	G6P dehydrogenase	Scott and Tatum, 1970
col-3	Structural	6PG dehydrogenase	Scott and Abramsky, 1973
col-10	Structural	6PG dehydrogenase	Scott and Abramsky, 1973
rg-1	Structural	Phosphoglucomutase	Brody and Tatum, 1967; Mishra and Tatum, 1970
rg-2	Structural	Phosphoglucomutase	Mishra and Tatum, 1970
?	?	Phosphohexoisomerase	Murayama, 1969

[a] G6P dehydrogenase, glucose-6-phosphate dehydrogenase; 6PG dehydrogenase,
6-phosphogluconate dehydrogenase.

pleiotropic effects on cellular metabolism considering the specialized role of the shunt in the formation of NADPH. NADPH is utilized by cells to provide reducing power for synthesis of cellular components, in particular fatty acids and sterols.

The enzymatic defects of the PMS mutants are of the K_m type where the full complement of activity is present under conditions of substrate saturation but the binding constants for the substrate and/or cofactor are reduced (Table IV). The adverse consequences from this type of change occur because *in vivo* metabolite concentrations are the rate-limiting factor in determining the velocity of enzyme-catalyzed reactions. Since the K_m values of most enzymes approximate the *in vivo* concentration of the substrate, reductions in substrate binding constants by two- to tenfold of a critical enzyme within a metabolic sequence can drastically reduce the efficiency of the pathway.

In addition to altered kinetic properties, the mutant dehydrogenases exhibit increased thermolabilities (Table IV) and most also have isoelectric points that differ from the wild-type enzyme (for references, see Table III). These changes are indicative of structural alterations and suggest that the mutant genes encode the primary sequences of the enzymes.

Studies of the biochemical lesions resulting in morphological changes can provide insight into the structure and regulation of the enzyme in

TABLE IV   Properties of the Wild-Type and Mutant Glucose-6-Phosphate and 6-Phosphogluconate Dehydrogenases[a]

Enzyme source	Half-life at 50°C (min)	K_m ($\times$ 10⁵) (M)		
		NADP	Glucose 6-phosphate	6-Phosphogluconate
Glucose-6-phosphate dehydrogenase				
Wild-type	9.0–11.0	1.3	2.9	
Mutants				
bal	2.8– 3.1	1.3	10.0	
fr	2.8– 3.1	1.3	10.0	
col-2	2.8– 3.5	2.1	10.0	
6-Phosphogluconate dehydrogenase				
Wild-type	20–30	1		3
col-1	54	1		11
col-3	5	3		9

[a] Data are compiled from Scott and Tatum (1970) and from Scott and Abramsky (1973).

question. This is illustrated by the glucose-6-phosphate dehydrogenase mutants (Scott and Tatum, 1970, 1971; Scott, 1971). The purified enzyme is a multimeric structure of molecular weight of 206,000 composed of four subunits with identical molecular weights (57,000). This subunit structure of the glucose-6-phosphate dehydrogenase suggests a molecular basis for complementation between the col-2, bal, and fr mutants and also for the multigenic control of the enzyme. Three isozymes of the Neurospora glucose-6-phosphate dehydrogenase are separable by electrofocusing. The available evidence indicates that each isozyme is composed of all three nonidentical subunits. A model for the structure of the glucose-6-phosphate dehydrogenase has been proposed based on this information (Scott and Mahoney, 1976). That the glucose-6-phosphate dehydrogenase is a complex molecule is not surprising considering that it catalyzes the first step of the PMS and occurs at a branch point of carbohydrate metabolism. The enzyme exhibits regulatory properties and is thought to be the control point for the PMS. NADP acts as a positive effector of glucose-6-phosphate dehydrogenase. The cofactor promotes the aggregation of the dimeric form of the enzyme to the tetrameric state with a concomitant increase in specific activity. Presumably, therefore, fluctuations in NADP levels regulate the activity of the PMS.

By examining the properties of the defective enzyme and the phenotypes of double mutants and heterokaryons constructed between nonallelic mutants carrying the same enzymatic lesion, a correlation is obvious between the degree of the phenotypic abnormality and the enzymatic defect. For example, the properties of the mutant enzyme are intermediate in heterokaryons and the phenotypes are wild-type-like, whereas both characteristics are more extreme in the double mutants than in the constituent single mutants. This relationship between the morphological and enzymatic defects suggests that the morphological aberrations are caused by the defective enzyme.

However, it is not immediately obvious how alterations in the PMS lead to morphological abnormalities. No gross changes in cell wall composition of the PMS mutants could be detected (W. A. Scott, unpublished observations) as found for the rg mutants which carry a defective phosphoglucomutase (Brody and Tatum, 1967; Mishra and Tatum, 1970). If the primary function of the PMS is the synthesis of NADPH, reduced levels of this pyridine nucleotide would be expected in the glucose-6-phosphate dehydrogenase and 6-phosphogluconate dehydrogenase mutants. Brody (1970, 1972) has found this to be the case with the col-2, bal, fr, and col-3 mutants (Table I). Although the pyridine nucleotide content of col-10 has not been measured, a similar reduction is presumed to occur in this mutant. The PMS defects also result in a decrease in the level of NADH and the oxidized pyridine nucleotides. Since other evidence suggests that decreases in NAD and NADH content have no adverse effects on the morphological characteristics of the organism (Brody, 1972), only the NADPH (and NADP) reduction is thought to have morphological consequences.

A second pleiotropic effect specific to the PMS defects is a reduction in linolenic acid content (Brody and Nyc, 1970; Nyc and Brody, 1971). A direct relationship exists between the level of linolenic acid and NADPH in the PMS mutants. This finding is consistent with the idea that the lower content of linolenic acid is a direct consequence of the NADPH deficiency. For this to be the case, the synthesis of linolenic acid must be more sensitive to NADPH levels of the cell than other unsaturated fatty acids. Supplementation of the PMS mutants with linolenic acid is one method whereby the functional significance of the polyunsaturated fatty acid reduction can be assessed. As described (Section II,C), addition of polyunsaturated fatty acids to agar cultures of *fr* cause the mutant to assume a wild-type-like phenotype and also increase the growth rate of the mutant threefold. No response, however, is obtained with fatty acids that contain a single double bond or with saturated fatty acids (Scott, 1976a). This positive response and the specificity of the response with respect to fatty acid structure are indicative of the fact that the linolenic acid deficiency contributes to the morphological abnormalities of *fr*. The importance of this observation is that it suggests plausible mechanisms whereby defects of the PMS can affect cellular ultrastructures. Defective membranes can plausibly lead to alterations in cell wall synthesis in addition to membrane-associated functions such as adenylate cyclase (Section II,C). However, the significance of the linolenic acid deficiency in the other PMS mutants is yet to be clarified (Scott, 1976b). None of the other PMS mutants (Table III) respond to exogenous polyunsaturated fatty acids and none have a cyclic AMP deficiency.

III. MUCOR

Many species of fungi grow in either a hyphal or yeastlike form depending on the environment. This capacity for a change in growth habit is known as dimorphism. The yeast—mycelial dimorphism is characteristic of many species of *Mucor*. The yeastlike growth occurs by the formation of spherical buds, whereas hyphal growth proceeds by apical elongation and branching. Dimorphism in *Mucor* can be readily effected by changes in growth conditions.

Considerable effort has been directed toward elucidating the environmental factors that control the growth habit in *Mucor*. In general, anaerobic conditions, which favor high glycolytic activity, promote yeastlike development, whereas aerobic conditions, which favor oxidative metabolism, promote hyphal development. The atmosphere of the culture is a major determinant of the growth form in *Mucor*. In a mixture of CO_2 and N_2, yeastlike growth is favored, but under an atmsphere of pure N_2 either growth form can predominate (Bartnicki-Garcia and Nickerson, 1962a,b; Bartnicki-Garcia, 1968), depending on the hexose concentration. Under certain experimental conditions, the yeastlike growth can also be maintained in aerobic cultures.

As yet, little is known about the molecular events of dimorphism. Recent experiments by Sypherd and co-workers (below), however, have firmly established that cyclic AMP plays a role in the control of morphogenesis in *Mucor* (Fig. 2). Larsen and Sypherd (1974) noted that there is strong correlation between the effects of respiratory metabolism on hyphal morphogenesis in *Mucor* and catabolite repression in *Escherichia coli*. The early steps of catabolite repression in *E. coli* are enhanced by aerobic conditions, but repressed by fermentation (Okinaka and Dobrogosz, 1967). Repression of *lac* operon expression by glucose can be reversed by growth under anaerobic conditions. This is thought to result in part from the increased cyclic AMP content of anaerobic cells (Aboud and

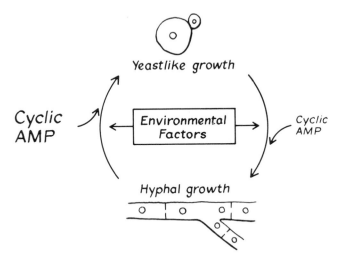

Fig. 2. Dimorphism in *Mucor*. Both environmental factors and changes in intracellular cyclic AMP content effect the reversible change between yeastlike and hyphal growth. It is likely that the influence of many of the environmental factors on the growth habit of *Mucor* are mediated via the cyclic AMP system of the organism, although this has yet to be demonstrated conclusively.

Burger, 1972). Proceeding on the assumption that cyclic AMP influences dimorphism in *Mucor,* Larsen and Sypherd (1974) examined the effects of exogenous cyclic AMP on the morphogenesis of yeastlike cells grown under 100% CO_2 to hyphae on exposure to air. Addition of the lipophilic derivative of cyclic AMP, dibutryl cyclic AMP, strongly inhibited the change to a hyphal growth pattern. Growth of cultures containing dibutryl cyclic AMP in air continued, albeit at reduced rates compared to the CO_2 cultures. The positive effects of exogenous dibutryl cyclic AMP on the hyphal to yeast transformation could also be demonstrated under aerobic culture conditions. Addition of dibutryl cyclic AMP to aerobic hyphae resulted in the appearance of budding yeastlike cells. The morphological changes are similar to those associated with the shift of aerobic cultures to an atmosphere of 100% CO_2. Measurements of endogenous cyclic AMP levels of yeastlike cultures prior to and after shifting from CO_2 to air indicated that a fourfold decrease in cyclic AMP content preceded the appearance of hyphal germ tubes. *Mucor* also contains cyclic GMP (Orlowski and Sypherd, 1976). In contrast to cyclic AMP, exogenous cyclic GMP and the dibutryl derivative of cyclic GMP had no effect on cell morphology and did not alter the influence of cyclic AMP on the hyphal–yeast dimorphism.

The above described effects of cyclic AMP are prevalent in cultures containing low (2%) glucose. Several observations suggest cyclic AMP alone is not sufficient for yeast formation. The inhibition of hyphal morphogenesis in CO_2 cultures on shifting to air or the induction of yeastlike morphogenesis in aerobic cultures, both of which are effected by exogenous dibutryl cyclic AMP, did not occur if glycerol, succinate, or maltose was substituted for glucose as the carbon source (Larsen and Sypherd, 1974; Paznokas and Sypherd, 1975). Since high glucose concentrations favor yeastlike development, it would appear that conditions which promote high glycolytic activity result in yeastlike growth. This conclusion is supported by the effects of various culture conditions on the growth habit of *Mucor* and further indicated that high cyclic AMP levels are correlated with glycolytic activity. Cyclic AMP levels of *Mucor* increase during the germination process regardless of the ultimate growth form, but the levels of the cyclic nucleotide are consistently higher under culture conditions leading to yeastlike growth (Paznokas and Sypherd, 1975). It seems plausible that cyclic AMP plays a central role in spore germination, and the cyclic AMP economy of *Mucor* is controlled by the cell physiology.

It has been possible to show that no correlation exists between the respiratory capacity and growth form of *Mucor* by manipulation of culture conditions. Respiration of the yeast form is high when grown in air and in the presence of dibutryl cyclic AMP. However, the respiration capacity of mycelia grown under N_2 is low. In contrast to *Saccharomyces,* the yeast form of *Mucor* contains mitochondria with developed cristae (Clark-Walker, 1973). Respiration in aerobic cultures is initially high and declines as if further development of respiratory capacity is lost. This suggests mitochondrial synthesis occurs predominately during the early stages of growth (up to midexponential growth) but can be modulated in response to environmental factors.

One of the most informative culture systems has been anaerobic growth under 100% N_2. The yeast–hyphal transformation in N_2 is readily reversible, and depends on the flow rate of N_2 over the culture media. The effect of N_2 flow rates is related to flushing the atmosphere over the culture. The most likely explanation of this phenomenon is that *Mucor* releases a volatile compound which influences morphogenesis (Mooney and Sypherd, 1976). Low flow rates of N_2 promote myelial development, while high flow rates favor yeastlike growth. As a result, accumulation of the volatile compound is thought to be required for mycelial growth. When the level of this compound drops below a critical level, yeastlike development occurs. It should be noted that this finding is not unusual in that extracellular products of many fungi are known to have morphogenetic effects (Gooday, 1974). In N_2 cultures, development of yeast and mycelia is less influenced by glucose levels than in CO_2 and aerobic cultures. Morphogenesis readily occurs at both high and low glucose concentrations in complex media, and higher levels of glucose are required for complete yeast transformation in semidefined media than in other culture conditions. Respiratory capacities of yeast and mycelial cultures in an N_2 atmosphere are similar and low. However, the differences in cyclic AMP content of the two growth forms are unchanged by growth in N_2. Since the major distinguishing feature between the yeast and mycelial forms of *Mucor* in N_2 cultures is the cyclic AMP content, it is of considerable interest to identify the volatile factor and determine whether it directly influences the cyclic AMP synthesis or breakdown. Because high concentrations of the factor are thought to favor hyphal growth, the factor should repress the synthesis of cyclic AMP or, alternately, stimulate the breakdown of cyclic nucleotide.

A respiration-deficient mutant of *M. bacilliformis* has been described (Storck and Morrill, 1971). Accompanying the loss of cytochrome oxidase activity in this strain is a loss of filamentation and sporulation. As a result, the mutant grows as a yeast in air. Additional mutants of *Mucor* unable to effect the dimorphoric change would greatly aid in sorting out the effects of the environmental factors on the physiology of the organism. These strains may also aid in deciding whether the dimorphic change is mediated solely by modulating cyclic AMP levels.

IV. *DICTYOSTELIUM*

A. Vegetative Growth

The cellular slime mold, *Dictyostelium discoideum,* has been employed by many investigators as a model system to study differentiation. Given an adequate food supply, the amoeboid cells undergo mitosis and exist as single cells. *Dictyostelium* can be grown in a defined axenic medium or fed bacteria which are ingested by phagocytosis (Loomis, 1975). Under optimal conditions, amoebae have been estimated to ingest up to five bacteria per minute, indicating that these cells are "professional" phagocytes. Unfortunately, little is known concerning the mechanisms of phagocytosis in this organism. Digestion is presumed to take place in food vacuoles which are extensive in amoebae feeding on bacteria. Phagocytizing cells are known to contain high levels of digestive enzymes and are also known to actively seek bacteria. Folic acid and possibly its degradation products are thought to be factors whereby the bacterial food supply is sensed (Pan *et al.,* 1972).

B. Genetic Analysis of Differentiation

The onset of differentiation in *Dictyostelium,* induced by starvation, consists of a complex series of events (Fig. 3). Mitosis stops, and the developmental sequence which ultimately results in the formation of the mature fruiting body or sorocarp is initiated. This involves migration of amoebae to central points, aggregation to form the slug or pseudoplasmodium, and finally the formation of the mature sorocarp which consists of the spore mass or sorus supported by a long, thin stalk. The available evidence suggests that differential gene expression is a feature of morphogenesis in *Dictyostelium.* Probably the most compelling evidence for this idea comes from the analysis of mutants. Loomis (1969b) isolated temperature-sensitive mutants of *D. discoideum.* These strains could be grouped into classes based on their phenotypes. Mutants of the first type were able to grow at 22°C but not at 27°C; however, differentiation proceeded normally at both temperatures. The second group of mutants grew at 22° and 27°C, but development of the sorocarp proceeded only at the lower temperature. Finally, the third set of mutants were unable to grow or differentiate at 27°C, while neither process was affected at 22°C. The mutations resulting in temperature-sensitive development appeared to affect cells only during the period of aggregation before the construction of the sorocarp. Mixing experiments further indicated that mutations resulting in the three different phenotypes are nonidentical and recessive.

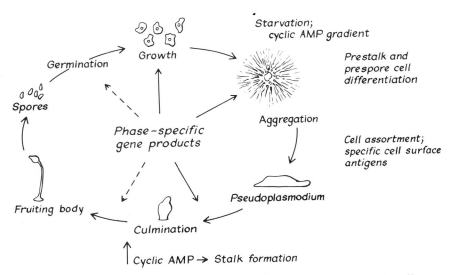

Fig. 3. Life cycle of *Dictyostelium*. Indicated wherever possible are the effectors and known molecular events associated with specific developmental steps. Phase-specific gene products, as delineated by mutant studies, are shown as solid lines. Mutational blocks affecting the first three steps of development have been recognized (Yamada *et al.*, 1973). Mutants affecting the fruiting body and spore morphology are also known (see Loomis, 1975). Therefore, it is likely that specific gene products are also required for the later differentiation events. These are designated as dashed lines since the specificities of these gene products are less certain.

These results suggest at least one gene or group of genes are specific to the growth phase of *D. discoideum,* one gene or group of genes specific to the developmental phase, and one gene or group of genes common to growth and development. Assuming differential gene expression is necessary for differentiation, three sets of genes of the type delineated by the temperature-sensitive mutants would have been predicted (Loomis, 1969b).

By screening 71 mutants of *D. purpureum* unable to carry out normal development, Yamada *et al.* (1973) found that these mutants also can be divided into genetically distinct classes. Mutants of the first type do not aggregate. The mutants of the second type form conical aggregates and stop. Those of the third type develop further to form finger-shaped vertical masses, the first stage of stalk development, but differentiation to fruiting bodies does not occur. As found with the temperature-sensitive mutants, complete synergism between mutants of phenotypically different classes occurred, and normal fruiting bodies were formed in mixed cultures. However, development did not proceed beyond the characteristic block in mixed cultures of phenotypically similar mutants. Yamada *et al.* (1973) consider these results to indicate that there are a minimum of three distinct developmental stages in fruiting body formation and that each is controlled by a different set of genes. This conclusion is compatible with results of Firtel (1972). From DNA–RNA hybridization experiments, he found that approximately half of the nonreiterated portion of the *Dictyostelium* genome is transcribed throughout the life cycle. Of this RNA, approximately half can be subdivided into stage-specific subclasses.

The synergistic behavior between mutants with nonidentical developmental blocks is of interest. Sussman and Lee (1955) observed that the synergistic response with another set of developmental mutants occurred only if the two nonidentical cell populations were physically associated and not when separated by thin membranes. This early observation was thought to indicate that cell–cell contact is essential for development. Yamada *et al.* (1973) postulated that either diffusable substances control development or diploid heterozygotes are formed in complementing populations.

By adding actinomycin D to block DNA-dependent RNA synthesis at various times after the onset of starvation in wild-type cultures, differentiation was terminated at three developmental stages identical to those delineated by developmentally aberrant mutants (Yamada *et al.*, 1973). The actinomycin D experiments suggested genes controlling development in *D. purpureum* are divided into three groups, which, by this criterion, were indistinguishable from those identified by mutation. The implication of these results is that there are three distinct developmental stages in fruiting body formation and coordinate expression of gene(s) are necessary for each.

In addition to the above described mutants, other morphological mutants have been isolated (Sussman and Sussman, 1953; Loomis and Ashworth, 1968). Drug resistant (Fukui and Takeuchi, 1971) and temporal (Sonneborn *et al.*, 1963; Loomis, 1970) mutants have also been described. However, formal genetic analysis in *Dictyostelium* has made little progress. A parasexual cycle is known (Loomis, 1969a), but the lack of a simple system for genetic analysis has been a major impediment in applying a biochemical–genetic approach to elucidating the mechanisms of differentiation in *Dictyostelium*. Macrocyst formation and germination have been described (Erdos *et al.*, 1972; Nickerson and Raper, 1973a,b). Recent experiments indicate that macrocysts represent a true sexual stage (Erdos *et al.*, 1973, 1975; Macinnes and Frances, 1974). Hopefully, techniques for routine genetic analysis will be forthcoming in the near future.

C. Cyclic AMP

Under starvation conditions, amoebae of *Dictyostelium discoideum* stream together by a chemotactic process to form multicellular aggregates. The major advance in understanding this process was the recognition that the migration of amoebae toward collection points is directed by cyclic AMP.

That a cyclic AMP gradient is required for aggregation of *D. discoideum* amoebae is firmly established (Bonner, 1971). On media containing uniformly high concentrations of cyclic AMP, amoebae migrate but differentiate into separate and often isolated stalk cells (Bonner, 1970). The molecular events that initiate chemotaxis are largely unknown. Indirect evidence suggests that increases in intracellular cyclic AMP levels promote aggregation competence (Klein and Brachet, 1975). Klein (1976) found that the expression of adenylate cyclase activity is developmentally regulated. Exponentially growing amoebae contain no demonstrable adenylate cyclase activity, but activity is detectable in cell extracts within 3 hr after removal of the food source and reaches maximal levels by 4 hr. As a result, the capacity for the synthesis of cyclic AMP follows the same time course as for amoebae to reach aggregation competence. Whether intracellular cyclic AMP levels are the major determinant which regulate the ability of amoebae to respond to cyclic AMP gradients remains to be determined.

Presumably a portion of the cyclic AMP initially synthesized in starving amoebae is excreted at least by a few cells in order to establish the extracellular cyclic AMP gradient. Theoretical considerations indicate that an instability in the gradient must exist in order to explain aggregation patterns. This is thought to be provided by an extracellular cyclic AMP phosphodiesterase (Chang, 1968). In addition, these cells excrete a protein inhibitor of the extracellular phosphodiesterase (Gerisch *et al.*, 1972). The cyclic AMP gradient responsible for the streaming of amoebae is considered to be a product of the [rhythmic] excretion of cyclic AMP, its destruction by the cyclic AMP phosphodiesterase, and modulation of the phosphodiesterase activity by the protein inhibitor. In support of the necessity for the pulsatile secretion of cyclic AMP is the finding that mutants of *D. discoideum* which continuously release cyclic AMP into the medium do not aggregate (Gerisch, 1971; Durston, 1974).

Aggregation competent amoebae respond to a pulse of cyclic AMP by increasing their intracellular levels of cyclic AMP and then excreting the cyclic nucleotide into the medium (Roos *et al.*, 1975; Gerisch and Wick, 1975; Shaffer, 1975). In this manner amplification and relay of the chemotactic signal is obtained. For aggregating amoebae to derive direction from a cyclic AMP gradient, the cells must detect the chemotactic agent. This implies the presence of a cyclic AMP receptor on the cell surface, transmission of the signal to the inner surface of the cell membrane, and destruction of the bound cyclic AMP. Experiments by Malchow and Gerisch (1974) indicate that this sequence of events may occur in *D. discoideum*. Increased cyclic AMP binding and increased cell-bound phosphodiesterase activity are features of aggregation competent amoebae. Both activities reach maximal levels after 4 hr of starvation or at the time of full aggregation competence. However, when amoebae enter into aggregates, the levels of cyclic AMP appear to decrease (Ashworth, 1974). During this terminal phase of aggregation, adenylate cyclase and phosphodiesterase activities are also arrested (Klein, 1976; Darmon and Klein, 1976).

Since the synthesis of cyclic AMP may influence aggregation competence as well as the quality of the chemotactic signal, adenylate cyclase is the obvious factor which couples chemotactic and differentiation responses of amoebae to extracellular cyclic AMP (Klein, 1976). For this to be the case, binding of extracellular cyclic AMP should result in a transitory stimulation of adenylate cyclase activity which has yet to be demonstrated.

D. Aggregation

Amoebae aggregate and ultimately differentiate into fruiting bodies which consist of two basic cell types: stalk cells and spores. The available evidence indicates the existence of heterogeneity in amoebae populations prior to the formation of aggregates (Maeda and Maeda, 1974). Streaming amoebae can be fractionated into light and heavy cells by isopycnic centrifugation. During slug formation, the light cells become localized in the anterior of the aggregate and differentiate into stalk cells. The heavy cells, localized in the posterior of the pseudoplasmodium, differentiate into spore cells. Two reports indicated that the anterior, prestalk cells contain more cyclic AMP than the posterior, prespore cells (Maeda and Maeda, 1974; Pan *et al.*, 1974). This observation is thought to indicate that cyclic AMP plays a determinative role, depending upon its concentration, in the decision whether cells become stalks or spores. In support of this view, is the previously mentioned finding of Bonner (1970) that high extracellular concentrations of cyclic AMP promote the differentiation of amoebae into stalk cells.

It has been known since the early work of Raper (1940) that the front one-third of the pseudoplasmodium forms the stalk of the fruiting body and the posterior two-thirds become spores. Since the above discussed results and those of Takeuchi (1963) suggest heterogeneity of amoebae prior to aggregation, assortment of cells is likely during slug formation. There is some controversy, however, as to whether cell assortment occurs after cohesiveness is fully developed, particularly during pseudo-plasmodial migration (Bonner and Adams, 1958; Bonner et al., 1971; Farnsworth and Wolpert, 1971; Raper, 1940).

E. Cell Surface Receptors and Aggregation

Cell–cell interactions are undoubtedly important for aggregation. The question of whether discreet molecular entities on the cell surface interact to promote aggregation has been probed with monovalent antibodies (Beug and Gerisch, 1969; Beug et al., 1970). Blockage of two independent classes of sites inhibits aggregation (Gregg, 1956; Sonneborn et al., 1964; Beug et al., 1973). One group of antigens, expressed by aggregation competent amoebae, mediates the typical end to end association of aggregating cells. The second class of sites, present in growing and aggregating amoebae, mediates side to side contact. Cell adhesions mediated by these latter sites are inhibited by EDTA, whereas the developmentally dependent binding is not. Although these results do not establish whether the antibody binding sites are identical with the membrane sites that function in cell adhesion, the finding that the EDTA-insensitive sites are absent in a nonaggregating mutant suggested that this may be the case (Beug et al., 1973).

Rosen et al. (1973) reported the isolation of a protein from D. discoideum which agglutinizes formalinized sheep red blood cells. The expression of this protein is correlated with aggregation competence, requires protein synthesis, and parallels the appearance of the cell surface antigens described by Beug et al. (1973). Characterization of the agglutination factor indicated that it is a carbohydrate-binding protein with a specificity for galactose monosaccharides. This finding is of interest because of the involvement of cell surface carbohydrates in cell–cell interactions in other systems (see Rosen et al., 1973, for references). Siu et al. (1976) confirmed these results. The latter authors determined that a single lectinlike protein with a molecular weight of 100,000 and composed of four subunits is expressed concomitantly with cell cohesion. Both RNA and protein synthesis are required for expression.

By labeling cell surface proteins using the [^{125}I]lactoperoxidase–glucose oxidase coupled system (Phillips and Morrison, 1971; Hubbard and Cohn, 1972) followed by electrophoresis of the solubilized proteins in sodium dodecyl sulfate, Smart and Hynes (1974) found that changes in several plasma membrane proteins of amoebae accompany aggregation competence. Weeks (1975) and Geltosky *et al.*, (1976) employed [^{125}I]concanavalin A (Con A) to examine the composition of glycoproteins associated with the cell surface. At least fifteen Con A binding proteins are evident on the surfaces of growing amoebae, and the intensity of an 150,-000 MW protein increases between 6 and 18 hr of development. The interesting point emerging from these studies is that aggregation is accompanied by relatively few changes in cell surface proteins. However, it should again be emphasized that although the molecular architecture of the cell surface undoubtedly is important to cellular recognition and adhesion, no proof that any of the described changes are essential to amoeboid aggregation has been demonstrated.

F. Phase-Specific Enzymes

A number of phase-specific enzymes for *Dictyostelium* have been described (see Loomis, 1975). Changes in metabolic emphasis during development presumably result in remodeling of cells and synthesis of new structures unique to different developmental stages. After formation of the pseudoplasmodium, the commitment of cells to either become stalks or spores is not irreversible until formation of the fruiting body is initiated. Individual cells obtained by disruption of the slug readily revert to the vegetative, amoeboid stage (Raper, 1940). However, if deposited on a solid substratum, cells reaggregate almost immediately and repeat within 2–3 hr the developmental sequence which may previously have required up to 19 hr (Newell *et al.*, 1971). Construction of the fruiting body then proceeds at a normal rate. The changes in content of some phase-specific enzymes are significantly affected by disruption of the developmental sequence. For example, a ninefold increase in UDPG pyrophosphorylase occurs after aggregation as the result of an increased rate of synthesis (Franke and Sussman, 1973). Disaggregation of the pseudoplasmodium results in an immediate cessation of pyrophosphorylase synthesis which does not resume as long as the cells are dissociated. On reaggregation, however, cells initiate a new round of pyrophosphorylase synthesis. These results indicate that cell–cell contact is important in controlling the metabolic events of *Dictyostelium*. As in other systems, the mechanisms of this regulation are obscure. However, the important conclusions derived from these studies are that specific biochemical changes appear to associate with discrete stages of development and that the transcription and translation of genes occurs in a highly specific manner. Mechanisms other than transcriptional regulation, however, can lead to phase-specific increases in enzyme activity (Killick and Wright, 1974).

Since the differentiation of *Dictyostelium* results from starvation, it is an adaptation to hard times. Deprivation of a food source indicates the organism must utilize its internal energy stores. Hence, catabolism is surely an important mechanism for the maintenance of the vital functions of the differentiating cell. In agreement with this idea, protein degradation and amino acid metabolism proceed throughout development and appear to be a major source of energy (Hames and Ashworth, 1974). However, protein degradation must of necessity be a selective process for cell survival. Sussman and Lovgren (1965) suggested that the loss of an enzyme activity during differentiation may be as crucial as the acquisition of an activity. Analysis of the expression of phase-specific enzymes has been carried out in a number of the existing morphological mutants (see Loomis, 1975). These studies have provided information as to the control of such activities, but not their physiological role. Whether most of these activities are essential for development remains to be determined. Expression of an enzyme activity at a given stage of development is not a sufficient criterion to assume the activity is an obligatory component of the developmental program. This point is well established from the recent results of Free and Loomis (1974). A mutant which lacks α-mannosidase 1, a phase-specific enzyme, was found to develop normally, indicating this activity is not essential for differentiation. The mutation in this strain (M-1) appeared to be specific for α-mannosidase 1 in that it did not block the accumulation of other developmentally regulated enzymes.

Definitive proof of the essential role of an enzyme in morphogenesis will require a biochemical–genetic approach to the problem. Although numerous morphological mutants of *Dictyostelium* are available, the primary enzymatic lesions of these strains have yet to be elucidated. The difficulty, of course, is to pinpoint the defective activity. Temperature-sensitive mutants such as those isolated by Loomis (1969b) should greatly facilitate this effort in that the temperature lability of the affected enzyme will aid in its identification. If selection procedures can be devised for different phase-specific enzymes, such as the one described by Free and Loomis (1974) for α-mannosidase 1, this would obviously alleviate one of the major problems of studying biochemical differentiation.

Which metabolic pathways are important to the development of *Dictyostelium* remain to be determined. However, reasonable guesses can be made as to a few which influence morphogenesis. Almost certainly mutations affecting key reactions to carbohydrate metabolism would lead to morphogenetic changes. This assumption is based on the known compositions of essential phase-specific structures. One of these is the slime sheath. If the slime sheath is broken, cells readily revert to the vegetative state on a bacterial lawn (Raper, 1940). Therefore, this structure is required for the integrity of the pseudoplasmodium and the continued development of the organism. Approximately 70% of the sheath is composed of polysaccharides (Loomis, 1975). Likewise, cells entering into the stalk during culmination vacuolize and lay down rigid walls comprised mainly of cellulose (Ashworth, 1974). The stalk supports the sorus and promotes the dispersal of spores by natural mechanisms. It seems obvious that mutants incapable of polysaccharide synthesis would not be able to construct either the slime sheath or stalk. Detailed chemical analyses of the slime sheath and stalk are not yet available. Knowledge of the structural components of phase-specific structures will undoubtedly provide information as to which biochemical pathways participate in the synthesis of each.

The second obvious pathway important to the development of *Dictyostelium* is amino acid catabolism. If current thoughts are correct in that a major portion of cellular energy during development is provided by the degradation of amino acids, blocks in these pathways will also affect development. It should be noted that enzymes of amino acid and carbohydrate metabolism are known to be stage specific (Loomis, 1975).

V. CONCLUSIONS

We can now appreciate the complexity of the biochemistry and genetics which control differentiation in lower eukaryotes. Two major points emerge from the systems considered above. The first is the prevalent role of cyclic AMP in morphogenetic events. This should not be too surprising because differentiation in lower eukaryotes is initiated by depletion of the food supply or other conditions which impose hardship. This requires that these organisms be able to assess their environment and metabolic states. Cyclic AMP is ideal for this purpose in that its synthesis occurs at the cell surface and is modulated by the type and quantity of carbohydrate available for consumption. However, to generalize and assume a central role for cyclic AMP in morphogenesis of lower eukaryotes is premature.

The second point is that an integrated study of development is not only necessary but undoubtedly the most fruitful approach. Concentration on one aspect of development, such as the biology, genetics, or biochemistry, causes the investigator to lose sight of the overall problem and poses severe limitations. The advantage of the systems described here is that an integrated study is feasible. For this reason, lower eukaryotes will continue to contribute to our thinking about the molecular mechanisms of development.

ACKNOWLEDGMENTS

The work carried out in the author's laboratory is supported by grants GM 21071 and GM 16224 from the National Institutes of Health and by a grant-in-aid from the Research Corporation. The author would like to express his appreciation to Mrs. Grace Silvestri for typing the manuscript.

REFERENCES

Aboud, M., and Burger, M. (1972). *J. Gen. Microbiol.* **71**, 311–318.
Ashworth, J. W. (1974). *Biochem. Cell Differ.* **9**, 7–34.
Barber, J. T., Srb, A. M., and Steward, F. C. (1969). *Dev. Biol.* **20**, 105–124.
Bartnicki-Garcia, S. (1968). *J. Bacteriol.* **96**, 1586–1594.
Bartnicki-Garcia, S., and Nickerson, W. J. (1962a). *J. Bacteriol.* **84**, 829–840.
Bartnicki-Garcia, S., and Nickerson, W. J. (1962b). *J. Bacteriol.* **84**, 841–858.
Beadle, G. W. (1944). *J. Biol. Chem.* **156**, 683–689.
Beckwith, J., and Rossow, P. (1974). *Annu. Rev. Genet.* **8**, 1–13.
Beug, H., and Gerisch, G. (1969). *Naturwissenschaften* **56**, 374.
Beug, H., Gerisch, G., Kempff, S., Riedel, V., and Cremer, G. (1970). *Exp. Cell Res.* **65**, 147–158.
Beug, H., Katz, F. E., and Gerisch, G. (1973). *J. Cell Biol.* **56**, 647–658.
Bonner, J. T. (1970). *Proc. Natl. Acad. Sci. U. S. A.* **65**, 110–113.
Bonner, J. T. (1971). *Annu. Rev. Microbiol.* **25**, 75–92.
Bonner, J. T., and Adams, M. S. (1958). *J. Embryol. Exp. Morphol.* **6**, 346–356.
Bonner, J. T., Sieja, T. W., and Hall, E. M. (1971). *J. Embryol. Exp. Morphol.* **25**, 457–465.
Brody, S. (1970). *J. Bacteriol.* **101**, 802–807.
Brody, S. (1972). *J. Biol. Chem.* **247**, 6013–6017.
Brody, S., and Nyc, J. F. (1970). *J. Bacteriol.* **104**, 780–786.
Brody, S., and Tatum, E. L. (1966). *Proc. Natl. Acad. Sci. U. S. A.* **56**, 1290–1297.
Brody, S., and Tatum, E. L. (1967). *Proc. Natl. Acad. Sci. U. S. A.* **58**, 923–930.
Chang, Y. Y. (1968). *Science* **161**, 57–59.
Clark-Walker, G. D. (1973). *J. Bacteriol.* **116**, 972–980.
Darmon, M., and Klein, C. (1976). *Biochem. J.* **154**, 743–750.
deTerra, N., and Tatum, E. L. (1963). *Am. J. Bot.* **50**, 669–677.
Durston, A. (1974). *Dev. Biol.* **37**, 225–235.
Erdos, G. W., Nickerson, A. W., and Raper, K. B. (1972). *Cytobiologie* **6**, 351–366.
Erdos, G. W., Raper, K. B., and Vogen, L. K. (1973). *Proc. Natl. Acad. Sci. U. S. A.* **70**, 1828–1830.
Erdos, G. W., Raper, K. B., and Vogen, L. K. (1975). *Proc. Natl. Acad. Sci. U. S. A.* **72**, 970–973.
Farnsworth, P., and Wolpert, L. (1971). *Nature (London)* **231**, 329–330.
Firtel, R. A. (1972). *J. Mol. Biol.* **66**, 363–367.
Franke, J., and Sussman, M. (1973). *J. Mol. Biol.* **81**, 173–185.
Free, S. J., and Loomis, W. F. (1974). *Biochimie* **56**, 1525–1528.
Fukui, Y., and Takeuchi, I. (1971). *J. Gen. Microbiol.* **67**, 307–317.
Garnjobst, L., and Tatum, E. L. (1967). *Genetics* **57**, 579–604.
Garnjobst, L., and Tatum, E. L. (1970). *Genetics* **66**, 281–290.
Geltosky, J. E., Siu, C.-H., and Lerner, R. A. (1976). *Cell* **8**, 391–396.

Gerisch, G. (1971). *Naturwissenschaften* **58**, 430–438.
Gerisch, G., and Wick, U. (1975). *Biochem. Biophys. Res. Commun.* **65**, 364–370.
Gerisch, G., Malchow, D., Riedel, V., Müller, E., and Every, M. (1972). *Nature (London), New Biol.* **235**, 90–92.
Gooday, G. W. (1972). *Annu. Rev. Biochem.* **43**, 35–49.
Gregg, G. (1956). *J. Gen. Physiol.* **39**, 813–820.
Hames, B. D., and Ashworth, J. M. (1974). *Biochem. J.* **142**, 301–316.
Hartwell, L. H., Culotti, J., Pringle, J. R., and Reid, B. J. (1974). *Science* **183**, 46–51.
Henry, S. A., Atkinson, K. D., Kolat, A. I., and Culbertson, M. R. (1977). *J. Bacteriol.* **130**, 472–484.
Hubbard, A. L., and Cohn, Z. A. (1972). *J. Cell Biol.* **55**, 390–405.
Hubbard, S. C., and Brody, S. (1975). *J. Biol. Chem.* **250**, 7173–7181.
Killick, K. A., and Wright, B. E. (1974). *Annu. Rev. Microbiol.* **28**, 139–162.
Klein, C. (1976). *FEBS Lett.* **68**, 125–128.
Klein, C., and Brachet, P. (1975). *Nature (London)* **254**, 432–434.
Kornberg, A. (1974). "DNA Synthesis." Freeman, San Francisco, California.
Larsen, A. D., and Sypherd, P. S. (1974). *J. Bacteriol.* **117**, 432–438.
Lester, H. E., and Gross, S. R. (1959). *Science* **129**, 572.
Levey, G. S. (1971a). *J. Biol. Chem.* **246**, 7405–7407.
Levey, G. S. (1971b). *Biochem. Biophys. Res. Commun.* **43**, 108–113.
Lie, K. B., and Nyc, J. F. (1962). *Biochim. Biophys. Acta* **57**, 341–347.
Loomis, W. F. (1969a). *J. Bacteriol.* **97**, 1149–1154.
Loomis, W. F. (1969b). *J. Bacteriol.* **99**, 65–69.
Loomis, W. F. (1970). *Exp. Cell Res.* **60**, 285–289.
Loomis, W. F. (1975). "Dictyostelium Discoideum: A Developmental System." Academic Press, New York.
Loomis, W. F., and Ashworth, J. M. (1968). *J. Gen. Microbiol.* **53**, 181–186.
Macinnes, M. A., and Frances, D. (1974). *Nature (London)* **251**, 321–324.
Maeda, Y., and Maeda, M. (1974). *Exp. Cell Res.* **84**, 88–94.
Mahadevan, P. R., and Tatum, E. L. (1965). *J. Bacteriol.* **90**, 1073–1081.
Malchow, D., and Gerisch, G. (1974). *Proc. Natl. Acad. Sci. U. S. A.* **71**, 2423–2427.
Mishra, N. C., and Tatum, E. L. (1970). *Proc. Natl. Acad. Sci. U. S. A.* **66**, 638–645.
Mitchell, R. H. (1975). *Biochim. Biophys. Acta* **415**, 81–147.
Mooney, D. T., and Sypherd, P. S. (1976). *J. Bacteriol.* **126**, 1266–1270.
Murayama, T. (1969). *Jpn. J. Genet.* **44**, 439 (abstr.).
Newell, P. C., Longlands, M., and Sussman, M. (1971). *J. Mol. Biol.* **58**, 541–554.
Nickerson, A. W., and Raper, K. B. (1973a). *Am. J. Bot.* **60**, 190–197.
Nickerson, A. W., and Raper, K. L. (1973b). *Am. J. Bot.* **60**, 247–254.
Nyc, J. F., and Brody, S. (1971). *J. Bacteriol.* **108**, 1310–1317.
Okinaka, R. T., and Dobrogosz, W. J. (1967). *Arch. Biochem. Biophys.* **120**, 451–453.
Orlowski, M., and Sypherd, P. S. (1976). *J. Bacteriol.* **125**, 1226–1228.
Pan, P., Hall, E. M., and Bonner, J. T. (1972). *Nature (London), New Biol.* **237**, 181–182.
Pan, P., Bonner, J. T., Wedner, H. J., and Parker, C. W. (1974). *Proc. Natl. Acad. Sci. U. S. A.* **71**, 1623–1625.
Paznokas, J. L., and Sypherd, P. S. (1975). *J. Bacteriol.* **124**, 134–139.
Phillips, D. R., and Morrison, M. (1971). *Biochemistry* **10**, 1766–1771.
Raper, K. B. (1940). *J. Elisha Mitchell Sci. Soc.* **56**, 241–282.
Robison, G. A., Butcher, R. W., and Sutherland, E. (1971). "Cyclic AMP." Academic Press, New York.
Roos, W., Nanjundiah, V., Malchow, D., and Gerisch, G. (1975). *FEBS Lett.* **53**, 139–143.

Rosen, S. D., Kafka, J. A., Simpson, D. L., and Barondes, S. H. (1973). *Proc. Natl. Acad. Sci. U. S. A.* **70,** 2554–2557.

Schmidt, J. C., and Brody, S. (1976). *Bacteriol. Rev.* **40,** 1–41.

Scott, W. A. (1971). *J. Biol. Chem.* **246,** 6353–6359.

Scott, W. A. (1976a). *Annu. Rev. Microbiol.* **30,** 85–104.

Scott, W. A. (1976b). *Proc. Natl. Acad. Sci. U. S. A.* **73,** 2995–2999.

Scott, W. A., and Abramsky, T. (1973). *J. Biol. Chem.* **248,** 3535–3541.

Scott, W. A., and Brody, S. (1973). *Biochem. Genet.* **10,** 285–295.

Scott, W. A., and Mahoney, E. (1976). *Curr. Top. Cell. Regul.* **10,** 205–236.

Scott, W. A., and Solomon, B. (1975). *J. Bacteriol.* **122,** 454–463.

Scott, W. A., and Tatum, E. L. (1970). *Proc. Natl. Acad. Sci. U. S. A.* **66,** 515–522.

Scott, W. A., and Tatum, E. L. (1971). *J. Biol. Chem.* **246,** 6347–6352.

Selitrennikoff, C. P., Nelson, R. E., and Siegel, R. W. (1974). *Genetics* **78,** 679–690.

Shaffer, B. M. (1975). *Nature (London)* **225,** 549–552.

Shapiro, D. L., and Marchesi, V. T. (1976). *J. Biol. Chem.* **252,** 508–517.

Shatkin, A. J., and Tatum, E. L. (1961). *Am. J. Bot.* **48,** 760–771.

Sherr, S. I. (1969). *Bacteriol. Proc.* p. 120 (abstr).

Siu, C.-H., Lerner, R. A., Ma, G., Firtel, R. A., and Loomis, W. F. (1976). *J. Mol. Biol.* **100,** 157–178.

Smart, J. E., and Hynes, R. O. (1974). *Nature (London)* **251,** 319–321.

Sonneborn, D., White, G. J., and Sussman, M. (1963). *Dev. Biol.* **7,** 79–93.

Sonneborn, D. R., Sussman, M., and Levine, L. (1964). *J. Bacteriol.* **87,** 1321–1329.

Storck, R., and Morrill, R. C. (1971). *Biochem. Genet.* **5,** 467–479.

Sussman, M., and Lee, F. (1955). *Proc. Natl. Acad. Sci. U. S. A.* **41,** 70–78.

Sussman, M., and Lovgren, N. (1965). *Exp. Cell Res.* **38,** 97–105.

Sussman, R. R., and Sussman, M. (1953). *Ann. N. Y. Acad. Sci.* **56,** 949–960.

Takeuchi, I. (1963). *Dev. Biol.* **8,** 1–26.

Tatum, E. L., Barratt, R. W., and Cutter, V. M. (1949). *Science* **109,** 509–511.

Terenzi, H. F., Flawia, M. M., Tellez-Inon, M. T., and Torres, H. N. (1976). *J. Bacteriol.* **126,** 91–99.

Torres, H. N., Flawia, M. M., Terenzi, H. F., and Tellez-Inon, M. T. (1975). *Adv. Cyclic Nucleotide Res.* **5,** 67–78.

Weeks, G. (1975). *J. Biol. Chem.* **250,** 6706–6710.

Wood, W. B., Edgar, R. S., King, J., Lielausis, I., and Henninger, M. (1968). *Fed. Proc. Fed. Am. Soc. Exp. Biol.* **27,** 1160–1166.

Wrathall, C. R., and Tatum, E. L. (1973). *J. Gen. Microbiol.* **78,** 139–153.

Yamada, T., Yanagisawa, K. O., Ono, H., and K. Yanagisawa, K. (1973). *Proc. Natl. Acad. Sci. U. S. A.* **70,** 2003–2005.

5

Molecular Bases of Cytoplasmic Male Sterility in Maize

C. S. LEVINGS III and D. R. PRING

I. INTRODUCTION

The objective of this chapter is to briefly discuss contemporary information and possible molecular bases of inheritance and expression of cy-

PHYSIOLOGICAL GENETICS
Copyright © 1979 by Academic Press, Inc.
All rights of reproduction in any form reserved.
ISBN 0-12-620980-4

toplasmic male sterility (cms) of maize (*Zea mays* L.). A more general treatment of the subject can be found in Duvick's (1965) classic review. Substantially new and revealing data have been obtained since 1965, and much of this was generated as a result of the maternally inherited disease susceptibility of the Texas or T source of cms. The T source of cms was widely used in the production of hybrid seed, but the epidemic of southern corn leaf blight (race T of *Bipolaris maydis*) in 1970 showed that cytoplasmic uniformity, even in the presence of nuclear diversity, could have inherent dangers. Thus cms of maize represents an extremely useful trait, but also a trait which exposed maize to the inevitable variability of plant pathogens.

The various sources of cms, with their differential fertility restoration requirements and disease reactions, as well as their strikingly different mitochondrial DNAs, offer unique experimental materials for investigations designed to elucidate nuclear–cytoplasmic interactions in higher plants.

II. THE CYTOPLASMIC MALE STERILITY TRAIT

A. Pollen Sterility

The maize plant has a terminal inflorescence which is called the tassel. Male spikelets (flowers) are the separate units of which the tassel is composed. At maturity a normal or fertile tassel exserts its anthers and pollen is shed. It is these pollen grains which contain the young male gametophyte. In its most drastic form, the tassel of a cytoplasmic male sterile plant, does not exsert anthers, and therefore, no pollen is shed. Sometimes deformed anthers, called "sticks," are exserted, but they contain only aborted pollen grains. Female fertility is unaffected by the cms trait.

B. Disease Susceptibility

One of the male sterile cytoplasms, *cms-T* has been found to be highly susceptible to two leaf diseases. Prior to 1970 *cms-T* was extensively used in hybrid maize production (Ullstrup, 1972). It has been estimated that over 85% of the maize grown in the United States at that time contained the T cytoplasm. A severe disease outbreak led to the discovery that maize containing the T cytoplasm was highly susceptible to southern corn leaf blight (*Bipolaris maydis* race T, formerly known as *Helminthosporium maydis*) and yellow leaf blight (*Phyllosticta maydis*), and its use in seed production was curtailed. Thus far susceptibility and male sterility have shown a complete association, but it is not known if this association is due to linkage or if one cytogene is responsible for both phenotypic traits.

C. Strict Maternal Transmission

Rhoades (1933) working with a Peruvian source of cytoplasm conducted a thorough analysis of the first cytoplasmic male sterile of maize. He demonstrated that the male sterility was due to factors contributed by the female parent and that the nuclear genes had no influence on the expression of the male sterility. This was shown by crossing the male sterile plants as females with a wide range of fertile males. In the F_1 and subsequent generations all progenies were male sterile. The Peruvian sterile has since been lost.

The cytoplasmic male steriles T, C, and S manifest strict maternal inheritance. These steriles have proved constant in series of backcrosses to many different fertile male strains. Since these steriles have been used commercially in hybrid seed production, strict maternal inheritance has been repeatedly verified by the backcrossing procedure used to convert fertile inbred lines to cytoplasmic male sterile versions of the lines. In these cases, genic transmission of cms has been excluded, because the substitution for all the chromosomes did not result in male fertility.

D. Nuclear Fertility Restoring Genes

Cytoplasmic male sterility can be countermanded by certain nuclear genes which are called restorers of fertility (*Rf*). These genes can restore full pollen fertility to male sterile cytoplasms. Each cms type is distinguished on the basis of the specific nuclear genes which restore pollen fertility (Duvick, 1965). Sets of inbred lines which contain the various combinations of *Rf* genes are commonly utilized for the identification of cms type (Beckett, 1971; Gracen and Grogan, 1974).

Two dominant genes *Rf* and *Rf2*, which are located on chromosomes 3 and 9, respectively, are known to be required for the restoration of pollen fertility in the T cytoplasm (Duvick, 1965). These loci act in a complementary fashion in restoring fertility in that plants with T cytoplasm and the genotype *Rf __ Rf2 __* are restored fertiles. However, *cms-T* plants with the genotypes *rf rf Rf2 __*, *Rf __ rf2 rf2* or *rf rf rf2 rf2* are male sterile (nonrestored). *cms-T* restoration is at the sporophytic level. This is demonstrated by the fact that all the pollen of a plant with the genotype *Rf rf Rf2 rf2* is fertile, although only one-quarter of the pollen grains from this same plant contain the restoring alleles *Rf* and *Rf2*. Restorer genes may have other functions. This was indicated in studies where plants of *cms-T* which had been fully restored to fertility showed susceptibility to pathotoxin T intermediate between normal and nonrestored *cms-T* plants (Watrud *et al.*, 1975b; Barratt and Flavell, 1975).

The S cytoplasm is restored to pollen fertility by a single gene, *Rf3*, which has been mapped on chromosome 2 (Laughnan and Gabay, 1975b). Pollen restoration is gametophytic in nature (Buchert, 1961), that is, the genotype of the pollen grain determines the phenotype with respect to fertility. For example, *cms-S* plants with the genotype *rf3 rf3* are completely male sterile, plants with the genotype *Rf3 Rf3* are fully male fertile, and plants which are heterozygous, *Rf3 rf3*, produce one-half fertile and one-half aborted pollen grains. Fertile pollen grains have the genotype *Rf3* while aborted grains are *rf3*.

Since *cms-C* has only recently been discovered, it has not been intensively studied. The gene(s) necessary for the restoration of *cms-C* have not yet been identified. However, it is known that they are different from those required by the T and S cytoplasm. Restoration of *cms-C* is at the sporophytic level like *cms-T*.

Variations have been observed in the degree of fertility restoration within cms types. This may suggest that minute diversity exists within each cms class. Alternatively, these variations in restoration could be due to diversity in the nuclear restorer genes. Modifying genes which affect the degree of fertility have long been recognized. The situation is complicated further by the fact that pollen fertility is greatly influenced by environmental conditions. Readers are referred to Duvick's (1965) treatment of this subject for a more detailed account.

III. LOCATION OF FACTORS CONTROLLING THE CYTOPLASMIC MALE STERILITY TRAIT

One important question regarding cms is the location of the factors coding for this trait. It has been firmly established that this trait is not inherited in a Mendelian fashion, therefore, it is not under the control of nuclear genes. Recent studies suggest that the factors responsible for cms in maize are located on the mitochondrial DNA (mtDNA). The evidence comes from several diverse types of studies, but all have suggested a common conclusion. Over the years it has been conjectured that cms might be controlled by the chloroplast DNA or viruses. Presently, there is no evidence to support either alternative, although studies with the unusual S cytoplasm suggest that an extraneous agent may be associated with the *cms-S*. In the sections to follow, studies which have implicated mtDNA as the carrier of the factors for cms will be reviewed. Three different cytoplasmic male sterile types, T, C, and S, have been used in these studies. Certain unique features associated with each sterile type have proven valuable in the investigations.

A. Restriction Endonuclease Fragment Analysis of Organelle DNAs

Recently restriction enzymes have proved to be very powerful tools for the investigation of DNAs (Nathans and Smith, 1975). Restriction enzymes are endonucleases which cleave double-stranded DNAs at sequence-specific sites. When a homogeneous double-stranded DNA is digested by a restriction enzyme, it is cleaved wherever it contains the sequence-specific cleavage sites. This produces a characteristic array of fragments. If the DNA is of small complexity, separation of the restriction fragments by gel electrophoresis generates a characteristic fragment pattern. Such patterns serve as a fingerprint of the original DNA molecule in a fashion analogous to the tryptic fingerprints of proteins. The technique is called restriction enzyme fragment analysis. Specifically, if the fragment patterns of two mtDNAs are contrasted, differences would be detected when caused by extensive deletions, translocations of restriction sites within the genome, or the creation of new restriction sites or deletion of existing ones. Importantly, small deletions, minor translocations, and point mutations would not generally be resolved if the alterations did not change the detectable number of restriction sites.

The mtDNA from maize with normal or fertile (N), T, C, and S cms cytoplasms have been studied by restriction enzyme fragment analysis (Levings and Pring, 1976, 1977; Pring and Levings, 1978). Four different restriction enzymes, *Hind*III, *Eco*RI, *Bam*I, and *Sal*I, were used in these analyses. Each restriction enzyme has a different specific cleavage site. When the fragment patterns from the four cytoplasms, N, T, C, and S, were compared, they were readily distinguishable irrespective of the endonuclease employed (e.g., Fig. 1). Therefore, the four cytoplasms, N, T, C, and S each possess unique mtDNAs.

The restriction patterns are very complex. For example, the endonuclease *Hind*III cleaved the mtDNA from N maize into more than 50 fragments, while *Bam*I produced more than 40. Most of the variation among cytoplasms occurred among high molecular weight fragments. Although the cytoplasms tested were characterized by bands peculiar to that cytoplasm, most bands were common to all cytoplasms. Finally, the mtDNAs

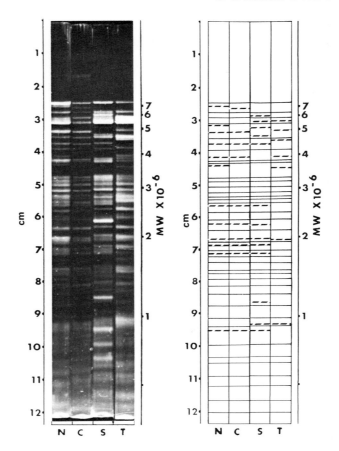

Fig. 1. Agarose gel electrophoretic patterns (left) and schematic (right) of *Hin*dIII digests of mitochondrial DNA from N (normal) cytoplasm and from the C, S, and T male sterile cytoplasms. Molecular weight estimates were from 0.5, 0.7, and 1% agarose gels; the patterns shown were from 1% tube gels electrophoresed for 17 hr at 1.9 V/cm. Dashed lines in schematic patterns indicate fragments that are not common to all cytoplasms.

from N, T, C, and S cytoplasms were studied in several different nuclear backgrounds. As expected, the restriction pattern distinctions were demonstrated to be consistent regardless of genetic backgrounds.

Another potential site for the cms trait is the chloroplast DNA (ctDNA). Therefore, restriction enzyme fragment analyses were carried out with the ctDNA from N, T, C, and S cytoplasms (Pring and Levings, 1978). Restriction patterns of ctDNA from N, T, and C cytoplasms were indistinguishable by *Hin*dIII, *Sal*I, or *Eco*RI endonuclease digestion. The ctDNA from the S cytoplasm, however, was slightly different from that of the other cytoplasms as indicated by a slight displacement of one band in *Hin*dIII digests.

We recently found that *Hae*III, which produces more than 55 fragments from ctDNA, will distinguish T cytoplasm ctDNA from the S, C, or N cytoplasms. The latter three cytoplasms were indistinguishable, while the T cytoplasm yielded one additional fragment. Thus very minor differences could be demonstrated among ctDNAs from the four cytoplasms.

The marked variation in mtDNAs, with the apparently minor variation among ctDNAs, represents circumstantial, but compelling evidence that the mtDNA is involved with the male sterility trait in maize. Recently, Quetier and Vedel (1977) have studied mtDNA and ctDNAs from normal (fertile) and cytoplasmic male sterile wheat by *Eco*RI restriction endonuclease fragment analysis. Distinctive fragment patterns were observed between the mtDNA from normal and male sterile cytoplasms, while the ctDNA studies revealed no differences. Thus their results with cms of wheat seem to be consistent with our findings with cms of maize. These results suggest that the association between cms and mtDNAs may be a general phenomenon among higher plants.

Presently, three distinctive cytoplasmic male sterile types are recognized in maize, T, C, and S. Identification has been based primarily on fertility ratings in different inbred backgrounds (e.g., Beckett, 1971). Other cytoplasmic steriles have been reported, but they have either been lost or not yet studied sufficiently to determine if they are unique. We have studied some newly isolated steriles by restriction enzyme fragment analysis of the mtDNA (C. S. Levings and D. R. Pring, unpublished data). In every case, they were identified as one of the three recognized cms types, T, C, or S. As yet, we have not unequivocally established that the new isolates of each cms type are genetically identical. Very small differences within a particular cms type could not necessarily be resolved by restriction enzyme fragment analysis. To date we have studied by restriction enzyme fragment analyses the mtDNAs from more than 25 steriles which had previously been identified by their fertility ratings in different inbred backgrounds. Without exception, the two techniques always identified the same cms type. The important conclusion is that each cms type is associated with a unique mtDNA. Undoubtedly alternative explanations are available; however, the most obvious one is that mtDNA sequence diversity present among the cms types is due at least in part to the factors responsible for the sterility trait.

The fragment patterns from *Hin*dIII-digested mtDNA of the various cytoplasms have been used to estimate the molecular weights (Pring and Levings, 1978). The complexity of these patterns and the uncertainty of masked bands precludes very accurate estimates. Nonetheless, the comparison of molecular weights of mtDNAs from the various cytoplasms is revealing (Table I). The mtDNA from normal cytoplasm had a minimum molecular weight of 131×10^6. The molecular weight of mtDNA from the T and C cms types were less than from normal cytoplasm. These results, in conjunction with a comparison of unique and common bands (Table I), suggest that cms types may be deficient for certain mtDNA sequences present in the fertile (N) cytoplasm. The possibility that the mtDNA of cms types is null for factors controlling male fertility must be seriously entertained.

A second line of evidence also suggests that the cms trait is not the result of a simple point mutation. The cms types T and C have never been reported to mutate from the male sterile to the male fertile condition (Duvick, 1965). In fact, their stability has been an important factor in their use in hybrid seed production. This is significant because mutations which result from deletional events would not generally be expected to revert. Furthermore, we continue to observe cytoplasm differences as additional endonucleases (*Sma*I, *Xho*I) are employed. Since each enzyme recognizes different restriction sites, it is reasonable to conclude that substantial differences exist among the mtDNA of the T, C, and S cms sources.

It is important in this context that small, but significant and repeatable, differences occur among normal cytoplasms of maize (Levings and Pring, 1977). These differences, which to date range from one to five new or missing bands in about 50 which are produced by *Hin*dIII, indicate that a small degree of mtDNA heterogeneity exists among inbreds of *Zea mays*. We interpret our collective data to indicate that the cms sources differ strikingly from normal cytoplasms, but that a low level of "background" heterogeneity is present in *Zea mays*.

TABLE I   Characteristics of *Hin*dIII Restriction Endonuclease Fragments of Mitochondrial DNA from Normal (N) Cytoplasm and the T, C, and S Sources of Male Sterile Cytoplasms of Maize

Cytoplasm	Number of bands			MW of common bands ($\times 10^6$)	N (%)	Minimum MW ($\times 10^6$)
	Total	Common with N	Unique			
N[a]	51	51	0	131	100	131
C	49	48	1	115	88	121
S	52	45	7	102	78	126
T	46	40	6	95	73	116

[a] B37 × NC236 or W64A.

Throughout these investigations, we have never obtained evidence of paternal inheritance of maize mtDNA and have observed no effect of the nuclear genome. Furthermore, it is apparent that mtDNA from the cms sources is stable and conserved. Texas cytoplasm mtDNA, for instance, was restricted from lines which had been backcrossed to normal paternal parents for at least 20 generations (Pring and Levings, 1978). These isolates, when compared to any other T source we have examined, always gave identical patterns. The level of sensitivity in detecting mixed cytoplasms probably ranges between 5 and 10%, which may not be very revealing if a low level of paternal transmission occurred. However, the similarity of all T cms mtDNA fragment patterns strongly suggests that paternally transmitted mitochondria, if present in progeny, are not expressed after repeated backcrossing.

B. Association of the T Source of cms with Disease Susceptibility

The T source of cms was widely used in the production of hybrid seed; Ullstrup (1972) estimated that about 85% of hybrid lines carried T cms in 1970. In the late 1960s, two fungal pathogens *Phyllosticta maydis* Arny and Nelson and race T of *Bipolaris maydis* (Nisikado) Shoemaker (formerly *Helminthosporium maydis* Nisikado and Miyake) were found to display preferential virulence on lines in T cytoplasm (see review by Ullstrup, 1972). Race T of *B. maydis* was especially important, causing the disastrous epidemic of southern corn leaf blight in 1970. This essentially unprecedented maternal inheritance of disease susceptibility, associated with T cms, had been described in the Philippines in the early 1960s but had not been observed in the United States. It was soon established that both fungal pathogens produce host-specific toxins (Hooker *et al.,* 1970; Comstock *et al.,* 1973; Yoder, 1973) and that these toxins preferentially affect the mitochondria of T cytoplasm, but not of N cytoplasm (Miller and Koeppe, 1971; Comstock *et al.,* 1973). Such evidence was suggestive of an association between mitochondria and disease susceptibility. The toxins of *B. maydis* cause swelling, uncoupling of oxidative phosphorylation, inhibition of malate/pyruvate or oxoglutarate oxidation, and increased oxidation of NADH (Miller and Koeppe, 1971; Peterson *et al.,* 1975). These responses are initiated within seconds of addition of the toxin to mitochondrial preparations. Evidence of *in vitro* reactions of T mitochondria to the toxin have been substantiated by *in vivo* physiological reactions, namely, an increase in respiration rate of detached leaf tissues (Bednarski *et al.,* 1977). Rapid (about 15 min) cytological effects of these toxins on mitochondria of root cap cells were also observed; swelling and internal disorganization were apparent in T, but not N, mitochondria (Aldrich *et al.*, 1977). Peterson *et al.* (1975) suggested the data indicated that the toxin interacts within the first ATP-coupled site of the electron transport chain, an inner membrane location. Inner membrane proteins are obvious candidates as mitochondrial gene products (Schatz and Mason, 1974).

Several additional characteristics of the T cms disease susceptibility interaction suggest a role of mitochondria in the expression of cms. Mitochondria isolated from Texas cytoplasm lines restored to fertility by the nuclear genes, *Rf*, *Rf2* exhibited a modified or attenuated reaction to the toxin (Watrud *et al.*, 1975b; Barratt and Flavell, 1975), and nuclear genotype, or source of the *Rf Rf2* genes, also influenced toxin reaction. These results suggest that fertility restoration genes alter the T mitochondria. Such an effect, that of a nuclear gene product on mitochondria, is not inconsistent with what is currently understood of fungal mitochondrial genetics (Schatz and Mason, 1974).

A unique type of evidence linking mitochondria to cms and disease susceptibility was recently described by Gengenbach and Green (1975), who used the toxins of race T of *B. maydis* to select resistant cells from T cytoplasm callus. Mitochondria isolated from the callus were unaffected by the toxin. Mature plants differentiated from this callus were resistant to the fungus and its toxin; mitochondria were similarly unaffected by the toxin, and the plants were fertile (Gengenbach *et al.*, 1977). Appropriate genetic analyses revealed that the alterations were cytoplasmic and not nuclear; crossing the plants as female to a nonrestoring nuclear genotype resulted in fertile progeny.

C. Identification of Possible Mitochondrial Gene Products Associated with cms

Several investigations have been designed to detect specific mitochondrial gene products which may be associated with cms and/or disease susceptibility. At present there is a paucity of knowledge concerning functional identification of proteins which are coded by higher plant mtDNA, although *in vivo* and *in vitro* comparative assays have offered evidence of discrete proteins which may be mitochondrial gene products (Leaver, 1976). Candidate proteins from studies of fungal and mammalian systems include constituent proteins of cytochrome oxidase, cytochrome *b*, and the oligomycin-sensitive ATPase (Schatz and Mason, 1974).

Evidence suggestive of an influence of T cms on cytochrome oxidase and cytochrome *b* content has been published (Pring, 1975; Watrud *et al.*, 1976). Barratt and Peterson (1977) were able to discern T cms-specific differences of chloroform–methanol-soluble proteins from submitochondrial particles and from partially purified mitochondrial ATPase complexes, indicating a potential correlation with data from other mitochondrial systems. Standard aqueous extraction techniques coupled with SDS–acrylamide gel electrophoresis or isoelectric focusing had generally not shown cytoplasmic differences (Bolens, 1975; Thornbury and Pring, 1976; Barratt and Peterson, 1977), although Watrud *et al.* (1975a) were able to observe slight variations between N and T inner and outer membrane proteins.

It would seem appropriate that studies designed to associate potential gene products with cytoplasms include materials of varying nuclear and cytoplasmic genotypes. Bolens (1975) was able to demonstrate proteins associated with mitochondria of specific inbreds, and the inclusion of the C and S cms sources (resistant to race T of *B. maydis*) would add cytoplasms with differential characteristics. Examination of the four cytoplasms in different nuclear backgrounds has recently indicated cytoplasm-specific mitochondrial proteins, but the site or function of these proteins is unknown (D. W. Thornbury and D. R. Pring, unpublished data). The differential sensitivity of the T, C, and S sources of cms to the toxins of *B. maydis* and *P. maydis* make these cytoplasms attractive model systems for the study of mitochondrial biogenesis and function in higher plants, since they provide materials of different genetic and physiological characteristics.

D. Cytological Evidence Implicating Mitochondria

A role of mitochondria in the expression of cms in maize has also been postulated from cytological studies of microsporogenesis. Mitochondrial degeneration was observed by the tetrad stage in the tapetum and middle layer of anthers from T cytoplasm but not N cytoplasm (Warmke and Lee, 1977). This degeneration was expressed as loss of cristae, internal disorganization, and swelling of the mitochondria. Plastids and other organelles did not differ in structure until late in anther development. Size and numbers of mitochondria or plastids were not different between N and T cytoplasm until after anther degeneration was evident (S. L. J. Lee and H. E. Warmke, personal communication). It is interesting to note, however, that a rapid increase (20- to 40-fold) in the numbers of mitochondria per cell occurred in the tapetum and sporogenous cells during meiosis of both the N and T lines. This dramatic increase, culminating at about the time when mitochondrial degeneration in T lines was first observed, led Warmke and Lee (1978) to propose that T mitochondria were unable to function competently under these stress conditions.

IV. THE S CYTOPLASM

Although the S cytoplasm confers male sterility just as T and C, it has other unusual attributes. Earlier, differences in the manner and genes required for fertility restoration were pointed out. Additional distinctions which are especially interesting have recently been discovered. These results have come from studies of the mtDNA of the S cytoplasm and from studies of mutations involving the S cytoplasm and its restorer.

A. Association with Unique Plasmid-like DNAs

The mtDNA from the S cytoplasm has been isolated and partially characterized (Pring *et al.*, 1977; Pring and Levings, 1978). In addition to the usual high molecular weight mtDNAs, the S cytoplasm contained two unique plasmid-like DNAs which were first identified by gel electrophoresis. These DNAs had molecular weights of 3.45×10^6 (S-F) and 4.10 (S-S) $\times 10^6$. We have tentatively demonstrated by electron microscopy that these molecules have a linear conformation. However, it may be that the linear molecules have resulted from the breakage of naturally occurring circular molecules.

The small unique DNAs associated with the S cytoplasm have not been observed in mtDNA preparation from normal (fertile), T, or C cytoplasms. This observation has been repeatedly verified among many sources of these cytoplasms. In addition, the unique DNAs are not present in mtDNA preparations from teosinte and tripsacum, which are close relatives of maize (C. S. Levings and D. R. Pring, unpublished). Conversely, these unique DNAs have been found in every S cytoplasm studied, regardless of source or nuclear background. This includes nine different sources of the S cytoplasm. These associations suggest a causal relationship between the unique DNAs and the S type of male sterility.

The unique DNAs associated with the S cytoplasm manifest strict maternal transmission. This fact was demonstrated in crosses where plants containing the S cytoplasm and the restorer gene, *Rf3*, were used as male parents in crosses with normal cytoplasm individuals. The *Rf3* gene restores pollen fertility to the S cytoplasm. In these test crosses, the unique DNAs associated with the S cytoplasm were not transmitted through the pollen. When S cytoplasm individuals were used as female parents, the unique DNAs were transmitted to the progenies irrespective of male parentage. Maternal transmission is expected of chloroplast and mitochondrial DNAs but not of nuclear DNAs. Finally, the unique DNAs have only been isolated from mitochondrial preparations. Efforts have failed to isolate these unique DNAs from chloroplast or nuclear preparations.

Recent electron microscopic investigations of the unique DNAs found in mitochondrial preparations from the S cytoplasm have revealed an unusual sequence arrangement (Levings *et al.*, 1978). When these DNAs are denatured and then briefly self-annealed, stem-loop configurations were observed by electron microscopy of formamide spreads (Fig. 2). This structure consists of a short double-stranded stem and a large single-stranded loop. The double-stranded stem contained 195 and 168 base pairs for the S-S and S-F DNAs, respectively. This constituted 3.1 and 3.2% of the S-S and S-F molecules, respectively. The occurrence of stem-loop configurations was interpreted as being due to the presence of terminal inverted repeats on the molecules. Intrastrand stem-loop structures form with first-order kinetics when the terminal inverted repeats pair (Broker *et al.*, 1977).

The interpretation of stem-loop structure was verified by several other studies. When S-S and S-F DNAs are denatured, briefly reannealed, and mounted for electron microscopy by the aqueous technique, stem-loop structures are not observed. Instead the molecules collapse and form bushlike structures. Bushlike structures are characteristic of single-stranded DNAs when mounted by the aqueous technique (Davis *et al.*, 1971). Since nearly 97% of the S-S and S-F reannealed molecules are single-stranded, collapsed molecules are expected.

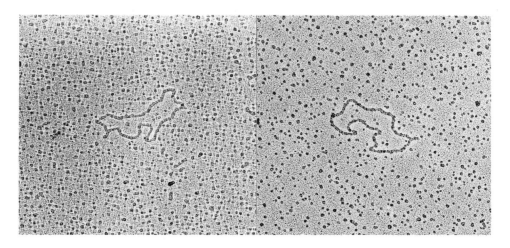

Fig. 2.   Electron micrographs of stem-loop configurations resulting from denaturing and briefly reannealing the S-S (left) and S-F (right) DNA molecules.

When the S-S or S-F DNAs were denatured, then self-annealed for 1 or 16 hr at 22°C, some homoduplexlike structures were observed. These structures contain double-stranded stems and "underwound loops" (Broker *et al.*, 1977). Underwound loops consist of single-stranded and double-stranded regions of variable lengths and distributions between the inverted duplication. They can occur when two complimentary strands each containing a stem-loop structure attempt to pair with each other. However, separated closed loops are topologically impenetrable, and no net interwinding is established. Limited annealing occurs which is a balance between the free energy of base-pair formation and the bond strains of uncharacteristically twisted strands. The occurrence of underwound loops further substantiates our conclusion that the S-S and S-F molecules contain terminal inverted repeats.

The importance of terminal inverted repeats on the unique DNAs from the S cytoplasm is not clear. Inverted repeats are often prominently involved with insertional events in lower organisms. It is tempting to hypothesize that the unique DNAs described here are in some way related to the unstable nature of the S cytoplasm.

B. Unstable Nature of the S Cytoplasm

In maize the cytoplasmic male sterile trait has proved to be remarkably stable. For instance, the T and S, especially T, male sterile cytoplasms have been employed in commercial seed production, and, therefore, the opportunity to observe changes from the male sterile to the fertile condition has been enormous. From these observations Duvick (1965) has concluded that these cytoplasms are notoriously stable. Although a few cases have been reported in which it was presumed that the cms trait had reverted to the fertile phenotype, rigorous substantiation of a heritable change at the cytoplasmic level has not been established. At any rate, such changes at the cytoplasmic level must be considered very rare events. The antithesis, change from the male fertile to the cytoplasmic male sterile condition, has been observed. However, the apparent changes to cms may have existed in these strains prior to discovery and appeared because of segregation of restorer genes. In this connection, attempts to produce new cms types by applying mutagens to strains with normal cytoplasms apparently have not been successful.

Several years ago Laughnan and his collaborators began a search for cases of changes in the S cytoplasm from the male sterile to male fertile condition (see Laughnan and Gabay, 1975b). Their search was successful in that they found mutations in which male steriles reverted to fertiles. Importantly, these studies were carried out under experimental conditions designed to elaborate the kinds of changes which occurred. Two types of changes were found: (1) cytoplasmic mutation from the male sterile to male fertile and (2) mutations giving rise to new nuclear restorer genes.

A brief description of the materials in which Laughnan found the mutations follows because of their possible relevance. Male sterile cytoplasm was introduced into seven sweet corn lines carrying the *shrunken-2* allele by backcrossing procedures. The sterile cytoplasm was the *Vg* source (dominant vestigial glume) whose pattern of restoration places it in the *cms-S* group. Therefore, they had male fertile maintainer and S male sterile versions of the seven *shrunken-2* inbred lines which have been maintained through 15 to 20 generations. The inbred lines involved were R839, R851, R853, R853N, R825, M825, and E1.

The crossing procedure which they used to identify the male fertile exceptions is as follows: When male sterile *cms-S* plants are crossed with maintainer pollen parents, the result F_1 progenies are expected to be male sterile except for the exceptional cases of male fertility. These male fertile plants may have wholly fertile tassels or fertile–sterile tassel sectors. In any event, they are crossed as pollen parents to *cms-S* male testers and either self-pollinated or crossed with a maintainer pollen parent. This technique determines if the newly arisen male fertile conditions are due to a change at the cytoplasmic level from S to normal, or alternatively, to a nuclear mutation at a restorer locus. Greater details concerning the technique are given in Laughnan's papers (Laughnan and Gabay, 1973, 1975a,b; Singh and Laughnan, 1972).

In the vast majority of cases, when the male fertile exceptions were analyzed by the above procedure, the male fertile trait was propagated through the female in crosses with maintainer plants, but was not transmitted through the pollen to test cross progeny. Therefore, it was impossible to explain the newly arisen male fertility on the basis of nuclear restorer genes, and it was concluded that the exceptional plants arose by a cytoplasmic change from a male sterile to a male fertile condition. Furthermore, the newly derived fertile cytoplasms have persisted through subsequent generations of propagation. At least 300 independently occurring cases of cytoplasmic changes from the S male sterile to male fertile have been identified and confirmed by the test procedure. Interestingly, most of these cases happened in the inbred line M825 which seems especially prone to the event. However, male fertile exceptions have been isolated from each of the other six inbred lines.

A number of male fertile exceptions have been identified whose test cross analyses indicated that the change did not occur at the cytoplasmic level. When these exceptions were crossed with *cms-S* male sterile testers, the progenies were all male semifertile, and these semifertile plants in subsequent tests exhibited a behavior expected of nuclear restorers of *cms-S*. Ten such new restorer strains have been isolated and partially studied. The ten arose either as fertile chimeras or fully fertile tassels.

The genetic behavior of the new restorers has been characterized in relation to *Rf3*, the naturally occurring restorer of *cms-S*. The mode of restoration of the ten new restorers is gametophytic as it is with the standard S restorer, *Rf3*. They differed from the standard *Rf3* locus in several respects. The new restorers have reduced transmission through the female gametophyte, a reduction in kernel size, and a lethality of the restorer homozygotes. These differences discredited the initial assumption that new restorers arose from mutations at the standard *Rf3* locus and, therefore, would have identical or at least very similar characteristics. Interestingly, one of the new restorers, designated *IV*, which unlike the others, originated in a maintainer plant, appears to be without adverse effects.

Perhaps the most interesting facts concerning the new restorer genes has come from mapping studies (Laughnan and Gabay, 1975b). The new restorer genes have been given Roman numeral designations. The standard restorer locus, *Rf3*, has been mapped on chromosome 2, probably in the long arm. The new restorers *I* and *VIII* are on chromosome 8; *IV* and *VIII* are on chromosome 3; and *IX* and *X* are on chromosome 1. Although the other four restorers, *II*, *III*, *V*, *VI*, have not been precisely mapped, they are known to not be allelic with *Rf3*. Consequently, it appears that each new restorer is located at a unique chromosomal site.

These findings have led the investigators to propose the existence of a male fertility element which has the characteristic of an episome (Laughnan and Gabay, 1975a,b). In bacterial systems, episomes have the capacity to be transposed from one site to another or to be entirely lost. The transposing phenomenon is suggested by the newly arisen restorer genes in that apparently fertility restoring elements have been integrated at different chromosomal sites. In this connection, they have speculated why kernels carrying the newly arisen restorer genes manifested deleterious side effects. They have suggested that the distinctive behavior of new restorers may be due to differences in integration sites in the chromosomes or in that all fertility elements may not be alike. The latter possibility is supported by the fact that restorer *IV*, the only one of the ten restorers that has been shown not to be associated with deleterious side effects, is the only one which arose in a fertile maintainer cytoplasm.

Although the male fertile exceptions involve either a change at the cytoplasmic level or in the nucleus, the investigators have argued that the two events have a common origin. This was indicated by the fact that the two kinds of male fertile exceptions have arisen in the same strains and in both cases are expressed initially as either wholly male fertile individuals or as fertile sterile tassel chimeras. According to this idea, they suggest that the male fertile element is fixed in the cytoplasm when a cytoplasmic change from male sterile to male fertile occurs. Alternatively, when the element is fixed in the nucleus, it behaves as a restorer strain.

The apparent limitation of the S-S and S-F DNAs to only members of the S group invites the hypothesis that these molecules are indeed the "fertility elements" described by Laughnan. While rigorous data unequivocally linking the S-S and S-F DNAs to these phenomena are not available, additional circumstantial evidence of an association is being obtained. Four exceptional lines, with characteristics of the cytoplasmic change (sterile to fertile cytoplasm) were prepared, and mitochondrial DNAs were examined before and after digestion with restriction endonucleases. In each case no evidence of the S-S and S-F DNAs was obtained, as if the elements had disappeared from the mitochondria. As pointed out by Laughnan (1973, 1975a), these fertile cytoplasms could have arisen from rare paternal transmission of the cytoplasm (i.e., normal mitochondria). If so, the restriction digest should resemble that of a normal mtDNA and not of the S cytoplasm mtDNA. In each case examined to date, the restriction digest was not that of N mtDNA (J. R. Laughnan, S. J. Gabay, C. S. Levings, and D. R. Pring, unpublished). It would thus appear that the behavior of the S-S and S-F DNAs is much like that expected of the postulated "fertility elements."

V. THOUGHTS ON THE CMS SYSTEM AND ITS ORIGIN

It has been demonstrated with the various cms types and their nuclear restorer genes that the male fertility trait is under the control of both nuclear and cytoplasmic genes in maize. The exact nature of this interaction is unknown. It seems possible that nuclear and cytoplasmic genomes may share genetic information associated with the male fertility trait. The relationship between cms and their nuclear fertility restorer genes has provoked speculation as to their origin (Harvey et al., 1972). For instance, it has been suggested that certain genes responsible for male fertility were initially located on organelle DNA, but that they have been transposed to a nuclear site through evolutionary processes. When a redundancy of genetic information exists between the organelle genome and the nuclear genome, a portion of this information at one site (organelle) could be lost without an adverse effect so long as the same information was retained by the other (nucleus). Although the reciprocal situation is possible, the nucleus seems to have established a dominant role in information coding. Since parallel evolution is not necessarily equal, the distribution of male fertility genes between the organelle and nuclear genomes need not be equivalent among all segments of the population. This then would account for the fact that several different cms types exist and that each cms type requires specific nuclear restorer genes to achieve male fertility. Hybridization among population members could result in organelle and nuclear genome combinations where the coded information would be absent in the organelle as well as the nucleus; this situation would explain the male sterile condition.

How well does the present evidence support the proposed origin of the cms trait? Current information is still too sketchy to substantiate this theory; however, certain results are supportive. Restriction endonuclease digestions (*Hin*dIII and *Sma*I) suggest that N mtDNA has a higher molecular weight than T mtDNA.

Authenticated changes from the cytoplasmic male sterile to the male fertile condition have not been reported in the T and C cytoplasms (Duvick, 1965). These facts suggest that the cms trait may be due to a deletion rather than a simple point mutation. Seemingly, the mitochondrial genome of a normal contains a larger compliment of the genes necessary for male fertility than their cytoplasmic male sterile counterpart. Furthermore, in this connection, the cms condition can be countermanded by the addition of certain dominant nuclear genes, the fertility restoring loci. These findings are in accord with the proposed theory and suggest that the theory is at least a reasonable model for future study.

Little is known about nuclear restorer genes except that they restore fertility. It is not clear how they substitute for cytoplasmic genes controlling male fertility. They may have evolved in the nucleus and created an alternative pathway to circumvent the cytoplasmic lesion. Alternatively they may simply be cytoplasmic genes which have been transposed to a nuclear site. This latter suggestion is supported by the studies of the S cytoplasm where an episomal event has been proposed to explain the newly arising nuclear restorer genes. However, it may be that the S case is unique and that it is not applicable to restorers of T and C.

VI. ORGANIZATION OF THE   MITOCHONDRIAL GENOME

We have implicated the mtDNA in coding, at least in part, for some of the factors responsible for male fertility. The mtDNA of higher plants are the largest of mtDNAs so far studied. They have molecular weights of

70×10^6 daltons or more. Animal mtDNAs are among the smallest having molecular weights from 10×10^6 to 12×10^6. The fungi and protozoa appear to have intermediate size DNAs ranging from 18×10^6 to 49×10^6 molecular weight (Borst, 1977). One of the yet unanswered questions of molecular genetics is what, if any, additional information is coded by the large mtDNAs of plants.

Presently, there seems to be some uncertainty about the organization and size of the mtDNAs of higher plants. Some of this may be due to wide variation among the different species. For example, pea mtDNA has been studied by electron microscopy and renaturation kinetics (Kolodner and Tewari, 1972). Electron microscopy studies have revealed a single molecular class which exists as a covalently closed circle and measures 30 μm in contour length. The molecular weight of these molecules is 70×10^6 and 74×10^6 as determined by electron microscopy and renaturation kinetics, respectively. These two values are in reasonable agreement. mtDNA from several plant species, potatoes, virginia creeper, cucumber, and wheat, have been investigated by French workers (Vedel and Quetier, 1974; Quetier and Vedel, 1977). Their results have suggested that mtDNA is composed of several different molecular classes. For instance, electron microscopic examinations have demonstrated that the molecular weight of virginia creeper, cucumber, or wheat mtDNA is 60×10^6 to 70×10^6 daltons. However, when the weight was estimated by sizing the fragments resulting from digestion with the restriction endonuclease EcoRI, molecular weights of 120×10^6 to 165×10^6 were found. The discrepancy in molecular weights led them to suggest that molecular heterogeneity occurred in the mtDNA. Several molecules of approximately the same size but with different sequence arrangements were postulated to explain the results.

Similarly studies of potato mtDNA have indicated molecular heterogeneity (Vedel and Quetier, 1974; Quetier and Vedel, 1977). In this case, molecular weights of 60×10^6 were found by electron microscopy, 90×10^6 by the sizing of fragments of endonuclease digestion, and 99×10^6 by renaturation kinetics. Again the discrepancy in molecular weights suggests the molecular heterogeneity phenomenon. Recently, we carried out an electron microscopic examination of soybean mtDNA (Synenki et al., 1978). Several different sizes of circular molecules were observed, thus extending the heterogeneity phenomenon to soybeans.

Electron microscopy studies of maize mtDNA have also identified molecular heterogeneity (Shah et al., 1976). These examinations revealed at least four distinct classes of circular molecules which had contour lengths of 16, 22, 30, and 43 μm and molecular weight of 36, 47, 66, and 93×10^6, respectively. The 22 μm circles occurred more often than the other sizes. The 43 μm class may be a dimer of the 22 μm class. This suggestion is based on the size which is approximately double that of the 22 μm class and because the 43 μm circles occur with a low frequency. In addition, minicircles, less than 5 μm in length, were observed. Minicircles have been observed in chloroplast DNA preparations; however, their importance is not clear. In any event, it is now clear that several classes of molecules can be visualized in maize mtDNA preparations.

The explanation of molecular heterogeneity in maize mtDNA is not clear, but several possibilities are available. The information of the mitochondrial genome may be encoded on more than one molecule in a manner analogous to the multiple chromosomes found in the nucleus. This would be in contrast to the situation with chloroplast DNAs where the entire genome is apparently found on a single molecule. Data similar to maize have been obtained with sorghum. The molecular weights of sorghum mtDNA ranges from 100×10^6 to 200×10^6, while that of ctDNA was about 80×10^6 (Pring *et al.*, 1979).

An alternative explanation is that there may be more than one kind of mitochondrion, and each one may contain its own unique genome. Another possibility is that companion DNAs are found associated with a main mtDNA molecule. This situation would be analogous to that of the bacterial chromosomes and their associated plasmids. This possibility has already been indicated by the unique plasmid-like DNA found in the S cytoplasm types. Oligomeric series, recombinational events, and differential amplification are all occurrences which could give rise to heterogeneity. Whatever the basis of this heterogeneity, it appears that mtDNA molecules, as isolated, yield relatively constant restriction patterns. When several T mtDNAs, for instance, are examined, the patterns and the relative fluorescence of each band seem to be constant. This observation suggests that an equilibrium has been obtained in the processes which results in the apparent heterogeneity. The stoichiometry within a restriction pattern has not been evaluated because of the great complexity and overlapping bands. Finally, the presence of alien contaminating DNAs cannot be wholly discounted, although it seems remote. It appears likely that more than one of the causes may be implicated in molecular heterogeneity phenomenon.

VII. SUMMARY

The male fertility system is unusual in that it results from the interaction of nuclear and cytoplasmic genes. This concept has developed from studies of the male sterile cytoplasms and their nuclear restorer genes. The cytoplasmic male steriles represent types which contain "mutations" of those cytogenes responsible for the male fertility trait. Especially pertinent is the fact that these cytoplasmic lesions can be corrected by nuclear genes, the *Rf* loci. This relationship implies the possibility that the same function may be coded in different genomes and that complementation may occur between the genetic systems, the nuclear and cytoplasmic. It further suggests that the cytoplasmic and nuclear genetic systems have evolved in concert with each other.

Evidence has been presented which indicates that the factors responsible for the cms trait are borne on the mitochondrial genome. Currently, the evidence is circumstantial, but it is persuasive because it has come from several different lines of investigation. The restriction enzyme fragment analyses of mtDNAs have consistently detected marked differences between normal and male sterile cytoplasms, while similar studies with ctDNAs have proved negative. The unusual organization of the mtDNA is in itself supportive. The large size of the mitochondrial genome and its apparent packaging among several molecules (molecular heterogeneity) gives it a realistic potential for contributing to the fertility trait. The S type of male sterility is a particular convincing example. In this case, the male sterility phenotype only occurs when the two unique DNAs (S-S and S-F) are present in the mitochondrion. When these unique DNAs are lost, the phenotype reverts to male fertility. Studies of the "target site" of the pathotoxin T have also implicated the mitochondrion. This indirect evidence is supportive of the contention that the cms trait is carried by the mitochondrial genome because of the absolute association found between the two traits, cms and disease susceptibility, in the T cytoplasm. Finally, differences have been found between the normal and cms types with respect to various mitochondrial proteins and enzymes. This then constitutes the evidence; final proof awaits the mapping of mitochondrial genes which are responsible for male fertility and sterility.

Investigations of the S type of cms have revealed this system to be unstable. Mutations, both at the cytoplasmic and nuclear level, have been reported which have been explained on the basis of an episomal event. The presence of "fertility elements" which can be transposed between cytoplasmic and nuclear sites has been proposed. These elements may be the unique DNAs found associated with the mitochondrion from *cms-S*. If these elements can indeed be shown to be capable of transposition, it may be important in clarifying the origin of the cms systems.

The organization of the mitochondrial genome of maize is puzzling. Studies have indicated that the mitochondrion contain more than one class of DNA molecules. Although several explanations for the molecular heterogeneity phenomenon have merit, the exact cause remains unknown. Apparently, there is a relationship between the heterogeneity and the S type of cms. In this case, *cms-S* is associated with the presence in the mitochondrion of the two unique DNAs (S-S and S-F). It seems apparent that a clarification of the organization of the mitochondrial genome will be important in understanding the cms system.

ACKNOWLEDGMENTS

Some of the studies reported herein were supported in part by grants from the National Science Foundation (No. PCM 76-09956) and Pioneer Hi-Bred International, Inc.

REFERENCES

Aldrich, H. C., Gracen, V. E., York, D., Earle, E. D., and Yoder, O. C. (1977). *Tissue & Cell* **9**, 167–178.
Barratt, D. H. P., and Flavell, R. B. (1975). *Theor. Appl. Genet.* **45**, 315–321.
Barratt, D. H. P., and Peterson, P. A. (1977). *Maydica* **22**, 1–8.
Beckett, J. B. (1971). *Crop Sci.* **11**, 724–727.
Bednarski, M. A., Scheffer, R. P., and Izawa, S. (1977). *Physiol. Plant Pathol.* **11**, 129–141.
Bolens, P. (1975). Ph.D. Thesis, University of Missouri, Columbia (Univ. Microfilms, International, Ann Arbor, Michigan).
Borst, P. (1977). *In* "International Cell Biology 1976–1977" (B. R. Brinkley and K. R. Porter, eds.), pp. 237–244. Rockefeller Univ. Press, New York.
Broker, T. R., Soll, L., and Chow, L. T. (1977). *J. Mol. Biol.* **113**, 579–589.
Buchert, J. G. (1961). *Proc. Natl. Acad. Sci. U. S. A.* **47**, 1436–1440.
Comstock, J. C., Martinson, C. A., and Gengenbach, B. G. (1973). *Phytopathology* **63**, 1357–1361.
Davis, R. W., Simon, M., and Davidson, N. (1971). *In* "Methods in Enzymology" (L. Grossman and K. Moldave, eds.), Vol. 21, pp. 413–428. Academic Press, New York.
Duvick, D. N. (1965). *Adv. Genet.* **13**, 1–56.
Gengenbach, B. G., and Green, C. E. (1975). *Crop Sci.* **15**, 645–649.
Gengenbach, B. G., Green, C. E., and Donovan, C. M. (1977). *Proc. Natl. Acad. Sci. U. S. A.* **74**, 5113–5117.
Gracen, V. E., and Grogan, C. O. (1974). *Agron. J.* **65**, 654–657.
Harvey, P. H., Levings, C. S., III, and Wernsman, E. R. (1972). *Adv. Agron.* **24**, 1–27.
Hooker, A. L., Smith, D. R., Lim, S. M., and Beckett, J. B. (1970). *Plant Dis. Rep.* **54**, 708–712.
Kolodner, R., and Tewari, K. K. (1972). *Proc. Natl. Acad. Sci. U. S. A.* **69**, 1830–1834.
Laughnan, J. R., and Gabay, S. J. (1973). *Theor. Appl. Genet.* **43**, 109–116.
Laughnan, J. R., and Gabay, S. J. (1975a). *In* "Genetics and Biogenesis of Mitochondria and Chloroplasts" (C. W. Birky, P. S. Perlman, and T. J. Byers, eds.), pp. 330–349. Ohio State Univ. Press, Columbus.
Laughnan, J. R., and Gabay, S. J. (1975b). *In* "International Maize Symposium: Genetics and Breeding" (D. B. Walden, ed.), pp. 427–446. Wiley, New York.
Leaver, C. J. (1976). *In* "Genetics and Biogenesis of Chloroplasts and Mitochondria" (T. Bucher *et al.*, eds.), pp. 779–782. North-Holland Publ., Amsterdam.

Lee, S. L. J., and Warmke, H. E. (1978). *J. Bot.* (to be published).

Levings, C. S., III, and Pring, D. R. (1976). *Science* **193,** 158–160.

Levings, C. S., III, and Pring, D. R. (1977). *J. Hered.* **68,** 350–354.

Levings, C. S., III, Hu, W. W. L., Timothy, D. H., and Pring, D. R. (1978). *Maize Genet. Coop. Newsletter* **52,** 96–98.

Miller, R. J., and Koeppe, D. E. (1971). *Science* **173,** 67–69.

Nathans, D., and Smith, H. O. (1975). *Annu. Rev. Biochem.* **44,** 273–293.

Peterson, P. A., Flavell, R. B., and Barratt, D. H. P. (1975). *Theor. Appl. Genet.* **45,** 309–314.

Pring, D. R. (1975). *Plant Physiol.* **55,** 203–206.

Pring, D. R., and Levings, C. S., III (1978). *Genetics* **89,** 121–136.

Pring, D. R., Levings, C. S., III, Hu, W. W. L., and Timothy, D. H. (1977). *Proc. Natl. Acad. Sci. U. S. A.* **74,** 2904–2908.

Pring, D. R., Conde, M. F., and Warmke, H. E. (1979). In preparation.

Quetier, F., and Vedel, F. (1977). *Nature (London)* **268,** 365–368.

Rhoades, M. M. (1933). *J. Genet.* **27,** 71–93.

Schatz, G., and Mason, T. L. (1974). *Annu. Rev. Biochem.* **43,** 51–87.

Shah, D. M., Levings, C. S., III, Hu, W. W. L., and Timothy, D. H. (1976). *Maize Genet. Coop. Newsletter* **50,** 94–95.

Singh, A., and Laughnan, J. R. (1972). *Genetics* **71,** 607–620.

Synenki, R. M., Levings, C. S., III, and Shah, D. M. (1978). *Plant Physiol.* **61,** 460–464.

Thornbury, D. W., and Pring, D. R. (1976). *Proc. Am. Phytopathol. Soc.* **3,** 323 (abstr.).

Ullstrup, A. J. (1972). *Annu. Rev. Phytopathol.* **10,** 37–50.

Vedel, F., and Quetier, F. (1974). *Biochim. Biophys. Acta* **340,** 374–387.

Warmke, H. E., and Lee, S. L. J. (1977). *J. Hered.* **68,** 213–222.

Warmke, H. E., and Lee, S. L. J. (1978). *Science* **200,** 561–563.

Watrud, L. S., Baldwin, J. K., Miller, R. J., and Koeppe, D. E. (1975a). *Plant Physiol.* **56,** 216–221.

Watrud, L. S., Hooker, A. L., and Koeppe, D. E. (1975b). *Phytopathology* **65,** 178–182.

Watrud, L. S., Laughnan, J. R., Gabany, S. J., and Koeppe, D. E. (1976). *Can. J. Bot.* **54,** 2718–2725.

Yoder, O. C. (1973). *Phytopathology* **63,** 1361–1366.

6

Somatic Cell Genetics of Higher Plants: Appraising the Application of Bacterial Systems to Higher Plant Cells Cultured *in Vitro*

DONNA PARKE and PETER S. CARLSON

PHYSIOLOGICAL GENETICS
Copyright © 1979 by Academic Press, Inc.
All rights of reproduction in any form reserved.
ISBN 0-12-620980-4

I. INTRODUCTION

Since the description of synthetic media that support the indefinite growth of plant tissues (Gautheret, 1939; Nobécourt, 1939; White, 1939), plant tissue culture has progressed to encompass not only the culture of plant tissue explants on agar-solidified media but also the proliferation of smaller units of cells in aerated liquid media (Nickell, 1956; Torrey and Reinert, 1961; Lamport, 1964). The regeneration of a whole plant from a single cell by Braun (1959) has been followed by the demonstration of the totipotency of a number of plant cells from different species (Vasil and Vasil, 1972; Murashige, 1974; Narayanaswamy, 1977). With the development of the *in vitro* culture of plant cells, researchers recognized the parallels between microbial growth and plant cell growth *in vitro* and the potentials for applying microbial selection techniques to obtain mutant strains of higher plant cells (Blakeley and Steward, 1964; Northcote, 1969; Street, 1973a; Chaleff and Carlson, 1974a). The temporal and spatial advantages of using 10^8 plant cells with a generation time measured in hours rather than 10^8 whole plants with a generation time of months as a population from which to recover mutant lines is readily apparent. *In vitro* cell culture makes it easier to control the environment of plant cells: selective agents affect the cell population relatively uniformly and environment–genotype interactions can be minimized. Selection of plant cell mutants has been recognized as being of potential importance in obtaining cultivars useful to agriculture and horticulture, in the production of secondary metabolites processed by the drug industry, and in the elucidation of basic problems in biology (Northcote, 1969; Nickell and Heinz, 1973; Chaleff and Carlson, 1974a; Carlson and Polacco, 1975).

Genetic and physiological attributes of an organism determine the strategy of a selection system. If somatic cell genetics of higher plants is to derive models from bacterial genetics, a juxtaposition of the properties of the two groups of organisms that are relevant to the application of selection systems should be made. In this chapter we compare the growth media and the culture of bacteria and plant cells *in vitro*. Differences in the culture of the two groups of organisms are considered with respect to the application of mutant selection screens. Following the description of their nutritional demands and culture, we discuss some of the attributes of bacteria and plant cells *in vitro*, limiting our discussion of plant inheritance to nuclear events. The time course of growth, the nature of haploid genomes, the stability of chromosomes, and the occurrence of phenotypic variants are covered. The distinctive features of plant cells that affect the development of mutant selection systems are emphasized. With this background, we explore the application of bacterial selections to some problems in plant biology. Our choice of pertinent topics is necessarily limited, but we hope that it will help to inform the microbiologist of the complexities of somatic cell genetics of higher plants and the neophyte botanist of microbial lore that may be relevant to developing selective screens for plant cells *in vitro*. Furthermore, it is hoped that this review will stimulate even deeper consideration of the extent to which plant somatic cell geneticists can look to bacterial systems for inspiration and guidance.

II. MODEL BACTERIAL AND PLANT GENETIC SYSTEMS

A. Bacterial Models

Although higher plant cells may have more in common with the fungi or the blue-green algae than with bacteria, we have chosen to compare them with the nonfilamentous bacteria, the Eubacteriales. The amenability of the nonfilamentous bacteria to various genetic and cultural manipulations has resulted in the development of a great variety of selection schemes. Major advances in bacterial genetics have been made with *Escherichia coli* K12; consequently, some people tend to equate bacterial genetics with *E. coli*. However, the physiological genetics of bacteria is enriched with information derived from representatives of different genera. Aside from a minimal number of relevant references to other systems, we find sufficient examples that illustrate diverse selection schemes among studies of two groups of gram-negative bacteria: the coliform bacteria and the fluorescent pseudomonads, soil bacteria. The advantages of these bacteria as objects for physiological genetic studies include their ability to grow rapidly with a generation time of under 30 min, to grow as a uniform suspension of single cells, and to form a discrete colony of cells from a single cell on agar-solidified medium. Concomitant with the latter characteristic is their amenability to replica plating.

B. Higher Plant Systems

The term "somatic cell genetics of higher plants" encompasses a staggering array of methodologies as diverse as pollen culture and protoplast culture. Although recognizing the potential significance of these approaches, we find that more information is currently available about cultured, intact cells derived from somatic plant tissues. In a model system for physiological genetic studies of higher plant cells, the following goals would have been achieved.

1. The establishment of a friable callus culture that can be subcultured indefinitely

2. The cultivation of cells as a fine, uniform suspension of rapidly growing, viable cells

3. The regeneration of entire plants from callus tissue explants

4. The regeneration of entire plants directly from cells in suspension culture or from callus derived from such cells

5. The isolation of haploid cell lines by anther or pollen culture or by other standard techniques whenever the recovery of recessive mutant cell lines is desired

6. The stabilization of chromosomes at the desired karyotype.

Each of the above steps should be considered in approaching a new system. For example, frequently primary callus cultures will regenerate shoots readily, whereas older callus or callus derived from cells that have been in a suspension culture are recalcitrant to regeneration. In some cases, a suspension culture can be obtained rapidly by sieving a uniform population of cells in a callus as was done successfully by Müller *et al.* (1977).

Implicit in this chapter is the assumption that steps (1)–(4) have been accomplished. Without a suspension culture, conditions are not uniform for chemically mediated selections, and without regeneration of whole plants, it is difficult to determine definitively whether a variant cell line was created by a genetic event or by a nongenetic alteration of phenotype (see Section IV,D). The reader is referred to recent reviews on the regeneration of various plants (Reinert, 1973; Murashige, 1974; Narayanaswamy, 1977). Many plant varieties that are amenable to steps (1)–(4) do not meet steps (5) and (6). In Section IV, the problems of obtaining true haploids [step (5)] and of maintaining a stable complement of chromosomes [step (6)], are discussed.

III. COMPARATIVE *IN VITRO* CULTURE OF REPRESENTATIVE BACTERIA AND PLANT CELLS

A. Growth Media

The formulation of media that support chemoheterotrophic growth of plant cells *in vitro* has been largely an empirical matter. One of the most commonly used media is that developed by Murashige and Skoog (1962) for the growth of tobacco callus. The concentrations of inorganic nutrients in the medium were carefully optimized to support a growth yield that was not stimulated further by nutrients in the ash of tobacco leaf extract. Additional optimization of the organic components of the medium was carried out by Linsmaier and Skoog (1965). The Murashige and Skoog medium was based on that of White (1943). White's medium and other media used in the early culture of plant cells often required the addition of numerous organic growth factors and complex substances. By providing a proper balance of inorganic nutrients, the Murashige and Skoog medium minimized the need for organic supplements. It is one of a number of synthetic plant media that support the growth of plant cells from diverse species and tissues. Comparisons of the chemical composition and use of different plant media have been made in a number of recent reviews (Murashige, 1973, 1974; Gamborg *et al.,* 1976).

The macronutrient composition and organic content of two modified White's media, both of which have been used to culture tobacco cells, are shown in Table I. In the same table, the macronutrient and organic composition of representative synthetic media used for the culture of some chemoheterotrophic bacteria is given. Differences are apparent in the concentrations of nitrogen and the source of carbon and energy, in the provision of growth factors and growth regulators, and in the inclusion of a buffer. The limiting nutrient in the bacterial and the plant media differs.

1. Nitrogen Source

The two plant media in Table I illustrate the range of nitrogen concentrations used in plant media. The M-1D medium contains a level of nitrogen that is less than the amount used in the bacterial media. The nitrogen content of the Linsmaier and Skoog (L—S) medium is four- to sixfold higher than that commonly used in bacterial media. Most plant media are similar to the L—S medium in containing levels of nitrogen in excess of 20 mM. The disparity in the concentration of nitrogen in bacterial and the L—S media is particularly striking in light of the fact that nitrogen accounts for approximately 14% of the dry weight of a bacterial cell (Stan-

TABLE I   Macronutrient and Organic Composition of Synthetic Media for Plant Cells and Bacteria[a]

	Plant cell media		Bacterial media	
	M-1D	L-S		
	(Filner, 1965)	(Linsmaier and Skoog, 1965)	(Hegeman, 1966a)	(Monod, 1942)
A. Macronutrients (mM)				
NH$_4^+$	—	20.6	15	9.3
NO$_3^-$	2.5	39.4	—	—
Total nitrogen	2.5	60.0	15	9.3
H$_2$PO$_4^-$ or HPO$_4^{2-}$	0.1	1.25	50	15
Ca^{2+}	0.85	3.0	0.46	0.09
Cl$^-$	0.9	6.0	0.9	0.18
K$^+$	1.7	20	25	15
Mg^{2+}	1.5	1.5	2.3	0.25
Na$^+$	2.9	0.2	50	9.0
SO$_4^{2-}$	2.9	1.6	9.9	0.28
B. Organic constituents (mM)				
Carbon and energy source	58.4[b]	88[b]	10–25	up to 2.8[c]
Nicotinic acid	0.004	—	—	—
Pyridoxine-HCl	0.0005	—	—	—
Thiamin-HCl	0.0003	0.001	—	—
Myoinositol	—	0.55	—	—
2,4-D[d]	0.0023	—	—	—
IAA[e]	—	0.0114	—	—
Kinetin	—	0.00014 or 0.00093	—	—
C. Final pH[f]	6.2	5.6	6.85	7.0

[a] The media support chemoheterotrophic growth of plant cells and nonfastidious bacteria, respectively.
[b] Sucrose.
[c] Sucrose or glucose.
[d] 2,4-D, 2,4-dichlorophenoxyacetic acid.
[e] IAA, 3-indoleacetic acid.
[f] Before autoclaving the medium.

ier *et al.*, 1976) while nitrogen accounts for 1.5 to 2.3% of the dry weight of various parts of a whole plant (Salisbury and Ross, 1969). Whereas the proportion of nitrogen in whole plant materials is extremely low, the proportion of nitrogen in cells cultured *in vitro* can be as high as 3–5% depending on the concentration of nitrogen in the medium (Welander, 1976). The low proportion of nitrogen in plant cells may be due to the contribution of the plant cell wall, which is high in carbon and oxygen, to the dry weight. The wall of a callus cell, for example, accounts for 40–46% of the dry weight of a cell (Northcote, 1969).

Escherichia coli and many other bacteria can utilize a number of different inorganic and organic nitrogenous compounds as sole nitrogen source (Stephenson, 1949). Bollard (1966) has demonstrated that an array of organic nitrogenous compounds can serve as sole nitrogen source for an aquatic higher plant. Studies with cultured plant cells have employed the following compounds as sole nitrogen source in addition to nitrate or ammonium nitrate: urea (Riker and Gutsche, 1948; Heimer and Filner, 1970; Behrend and Mateles, 1975; Polacco, 1976), a complete amino acid mixture (Filner, 1966), certain single amino acids (Riker and Gutsche, 1948; Frank *et al.*, 1951), and ammonium if it is accompanied by succinate or a tricarboxylic acid cycle intermediate (Gamborg and Shyluk, 1970; Behrend and Mateles, 1976).

2. Source of Carbon and Energy

The most commonly employed source of carbon and energy for plant cells is sucrose. The concentration of sucrose is usually 58.4 to 117 mM (2 to 4%). Glucose and, to a lesser extent, fructose can also serve as carbon sources for plant cells (Gautheret, 1955; Gamborg *et al.*, 1976). Other sugars, alcohols, and organic acids are assimilated slowly or not at all, and some of these compounds actually inhibit growth with sucrose (Gautheret, 1955). Bacterial media, on the other hand, generally contain a carbon source at a level of 5 to 30 mM. Many species of bacteria can utilize a wide range of carbon sources. For example, *E. coli* strains can utilize approximately 20 organic acids, carbohydrates, and alcohols out of 118 compounds (Stephenson, 1949). Various *Pseudomonas* species can utilize from 11 to 108 different organic compounds out of 146 tested as growth substrates (Stanier *et al.*, 1966). The fastidious nature of plant cells with respect to carbon and energy sources is likely to influence the selection of mutants altered in the catabolism of organic compounds (Section V,A,3).

3. Growth Factors and Growth Regulators

The vitamin thiamin-HCl appears to be universally beneficial to plant cells *in vitro* and is incorporated into most media. Cell lines derived from particular species or tissues may require other vitamins or growth factors such as inositol (Dougall, 1972); in this respect, plant cell lines are similar to different bacterial species and strains. Unlike bacteria, however, plant cells usually require hormones in order to maintain continued growth *in vitro*. The two classes of hormones most essential to continuous cell division (Das *et al.*, 1956; Jouanneau and Péaud-Lenoël, 1967), DNA synthesis (Patau *et al.*, 1957), and mitosis (Das *et al.*, 1956) are the auxins and the cytokinins. The endogenous levels of auxin and cytokinin in cultured cells vary according to their rates of biosynthesis and degradation in addition to the balance between the sequestering of the hormones within the cells and the leakage of them from the cells. An auxin is supplied to most plant tissue cultures, while a cytokinin is provided if necessary. Different plant species, different varieties of a single species, and cell lines derived from different tissues of a plant may require different concentrations or types of auxin and cytokinin for continuous growth and cell division.

Plant hormones have a number of effects on plant cells and tissues. For example, they are known to stimulate an increase in certain enzymatic activities (Varner and Ho, 1976). The concentration of cytokinin that was supplied to cells was reported to influence the growth rate of tobacco callus (Helgeson *et al.*, 1969). It influenced the length of the lag phase, but not the growth rate, in suspension cultures of sycamore cells (Mackenzie *et al.*, 1972). Thus, a variable that is absent in the culture of bacteria must be considered in the culture of plant cells.

4. Use of Buffers

The absence of a buffer in plant media is in contrast with the bacterial growth media of Table I, which contain phosphate as a buffer in the pH range of 6.4 to 7.2. The amount of buffer needed depends on the concentration of carbon source supplied. The metabolism of 1 mole of glucose by bacteria, for example, may produce 2 moles of acetate, or acid. The initial pH of most plant cell media is 5.5 to 5.8, too low for phosphate to be an effective buffer; moreover, high levels of phosphate have been reported to be toxic to cells (Murashige and Skoog, 1962). Growth of plant cells does not lead to adverse changes in the pH of media that support cellular proliferation (Table II). When a highly homogeneous cell line of tobacco was cultured in media containing four different nitrogen sources, a deleterious pH shift occurred only in the medium containing ammonium chloride as sole nitrogen source (Table II).

Although there is not a deleterious change in the pH of nonselective plant culture media, there may be one in selective media. Unfavorable conditions for growth may cause the pH to shift excessively. The change in pH might lead to the failure to recover mutant cell lines. Monitoring and adjusting the pH of culture media may be crucial to the success of some mutant selection schemes.

TABLE II   The Influence of Nitrogen Source on the Final pH of Culture Media Used for the Growth of Cells of *Nicotiana tabacum* var. Wisconsin 38[a]

Form of nitrogen	Concentration of nitrogen (mM)	Final V_{30}/initial V_{30}	Final pH
A. KNO₃	18.8	5.0	5.7
NH₄NO₃	20.6		
B. NH₄Cl	20.0	5.7	4.9
Potassium succinate	10.0		
C. KNO₃	20.0	6.3	5.3
D. NH₄Cl	20.0	0.7	4.3
KCl	20.0		

[a] A highly homogeneous tobacco cell line that was habituated with respect to cytokinin was cultured on liquid Linsmaier and Skoog (1965) (L-S) medium containing the standard nitrogen source (A) and three alternative nitrogen sources (B, C, or D). The L-S medium incorporated 1.7×10^{-5} M 3-indoleacetic acid and no cytokinin. The initial pH of all media was adjusted to 6.0 with KOH before autoclaving. V_{30} designates the settled cell volume measured after allowing the cells to settle for 30 min. Flasks containing 50 ml of medium were inoculated with an average of 6 ml (V_{30}) of cells; the cells had been starved for nitrogen. One week intervened between inoculation and determination of final pH and volume. Experiments were conducted in triplicate and were repeated for verification.

5. Limiting Nutrients

The limiting nutrient in the bacterial and plant media differs. Generally, chemoheterotrophic bacterial growth is limited by the principal carbon source (e.g., see Hegeman, 1966a). Growth of plant cells in the M-1D medium is limited by the supply of potassium nitrate (Filner, 1966), while growth in the L—S medium has been proposed to be limited by the supply of water (Murashige and Skoog, 1962) or of nitrogen (Okamura *et al.*, 1975).

B. Culture of Cells in Agar-Solidified Media and Liquid Media

Escherichia coli or *Pseudomonas* cells can be plated at a density as low as one cell per petri dish. A density of 50 to 100 cells per dish is commonly used to obtain colonies for replica plating. Confluent growth represents on the order of 10^9 bacteria per petri dish, which is a density of approximately 4×10^7 cells per milliliter of medium. Similarly, liquid cultures can be initiated with less than a cell per milliliter, and the density may reach 10^9 to 10^{11} cells per milliliter depending on the culture medium and the conditions of growth.

Plant cells require a minimum density to survive in solid or liquid medium. The minimum cell density required depends on the particular plant cell line being used (Gautheret, 1942) and on the culture medium (Blakely and Steward, 1964). For example, the provision of optimal levels of auxin (Duhamet, 1953) or of cytokinin (Stuart and Street, 1971; Mackenzie *et al.*, 1972) lowered the minimum inoculum size required to get good growth in synthetic media by a factor of at least ten.

Dispersed plant cells can be plated relatively uniformly within the agar medium or on the surface of the agar. The Bergmann technique, in which dispersed plant cells are embedded in an agar matrix, has employed an inoculum density as low as 200 cells per dish (Bergmann, 1960). The simpler culture method of plating cells on the surface of the agar is useful where it is not important that the cellular units be separated. In order to have the plant cells dispersed relatively uniformly over a dish, a tobacco cell density of approximately 10^6 cells per 9-cm (diameter) petri dish, or 4×10^4 cells per milliliter of medium, was used in our laboratory.

In liquid medium, the minimum inoculation density was reported to be approximately 1.5×10^4 cells per milliliter for a sycamore suspension culture (Mackenzie *et al.*, 1972). In our laboratory, a cultured cell line of *Nicotiana tabacum* var. Wisconsin 38, characterized by its ability to be pipetted, required an initial density of approximately 2×10^4 cells per milliliter; the cells had been rinsed thoroughly with fresh medium prior to transfer. The suspended sycamore cells reached a final density of 2×10^6 cells per milliliter on a complex medium (Henshaw *et al.*, 1966); the tobacco cells reached a final density of 5×10^5 cells per milliliter on the synthetic Linsmaier and Skoog (1965) medium.

The large size of plant cells (12 to 300 μm in diameter) as compared with most bacteria (roughly 1 μm in diameter) is reflected in the large ratio of the settled cell volume of plant cells to the total volume of the medium plus cells. For the cell line of *N. tabacum* mentioned above, the ratio of the settled cell volume to the total volume of the medium plus cells was 0.03 upon inoculation and reached 0.8 or greater at stationary phase. The partial cell volume of bacteria in their culture medium is considerably less (Luria, 1960).

For carrying out a mutant selection, a plant cell suspension less dense than that attained by stationary phase is advisable. The maximum cell density attained by plant cells is on the order of 10^3 times less than that reached by bacterial cells. As a result, the screening for mutant plant cells on selective media and the estimation of reversion frequencies demand a much greater use of space and media than is required in work with *E. coli*. In addition, the large cell volume and required minimal inoculation density constrain the number of generations of cells that can be produced in a flask. Thus, growth of plant cells in a flask of liquid medium encompasses 5 to 7 cell generations maximally, in contrast to as many as 30 or more for an *E. coli* batch culture inoculated with 10 cells/ml.

C. Cloning of Cells

Implicit in the necessity for a minimum initial density is the fact that the cloning of plant cells under *nonselective conditions* is not a routine procedure as it is with many bacteria. Almost no plant suspension cultures are composed mainly of single cells. Frequently, a large proportion of the single cells observed are gigantic polyploid cells. A number of plant cell lines grow as elongated, filamentous cells in suspension (Bergmann, 1960; Blakely and Steward, 1964; Filner, 1965). The clonal derivation of cells that grow as a linear chain in a filament is clear. Other plant cell suspensions grow as clumps of cells, however, and the first step in cloning becomes the dispersal of cell clumps. Methods of separating cells include agitating the cells for a month in a medium containing an elevated sucrose concentration of 4% (Binns and Meins, 1973), exposing suspension cultures to cell wall degrading enzymes (Street, 1973b) or isolating protoplasts (Cocking, 1972). The effect of such treatments on the physiological and chromosomal stability of the cells may be deleterious. In addition, the use of pectinase has not proved useful in separating the cells of all cell cultures (e.g., see Jouanneau and Péaud-Lenoël, 1967). With or without treatments to separate cells, filtering and microscopic examination are essential to ensure the clonal nature of colonies that develop.

Once single cells or filaments of cells have been isolated, they can be cloned by the technically meticulous methods employing nurse callus tissue (Muir *et al.*, 1954, 1958) or a microchamber (Jones *et al.*, 1960). The Bergmann technique of plating described above is a more practical approach to cloning. Using this method, a number of researchers have found that the plating efficiency of cultured plant cells is extremely low. At an initial density that permitted units of single cells and small clumps of cells to grow as discrete colonies, the plating efficiency on complex media was 1 to 10% (Blakely and Steward, 1964), up to 20% (Bergmann, 1960), and up to 60% (Binns and Meins, 1973). Binns and Meins used the conditions developed by Gibbs and Dougall (1965) to achieve such unusually high plating efficiencies. With a synthetic medium, much variation in the plating efficiency was observed, but the values generally fell between 10 and 20% (Earle and Torrey, 1965). Calculating with the best values of Bergmann (1960), a researcher might expect to recover 40 discrete colonies per plate. Increasing the density or size of the cellular units increases the plating efficiency (Blakely and Steward, 1964), but the colonies tend to become confluent.

The use of a medium that was conditioned by other cells permitted the minimum initial cell density to be lowered tenfold (Stuart and Street, 1971). Two laboratories (Stuart and Street, 1971; Vysotskaya and Gamburg, 1975) have found that the conditioning factor is one or a number of low molecular weight compounds. Beyond that characteristic there is evidence that the conditioning factor is a thermostable, nongaseous substance (Vysotskaya and Gamburg, 1975), and there is contradictory evidence that at least one component of conditioning has a volatile nature (Stuart and Street, 1971; Dorée et al., 1972). Thus, in some laboratories at least, conditioned medium is unstable in its growth-promoting quality.

Another promising approach to plant cell cloning is the use of X-irradiated feeder cell layers, analogous to the cloning method used with mammalian cell cultures (Puck and Marcus, 1955). The use of X-irradiated protoplasts as feeder cells permitted the minimum initial inoculation density of protoplasts to be lowered by a factor of 100 and increased the cloning efficiency (Raveh et al., 1973).

In summary, separating some plant cell lines into single cells is difficult, and the plating efficiencies obtained with plant cells tend to be low and variable. These factors make plant cell cloning a rate-limiting step for a number of mutant selection procedures. For example, where cloning of cells is routine, it is feasible to isolate mutant cells from wild-type cells in a mixed population arising from an enrichment type of selection (Sections V,A,C).

D. Determination of Viable Cell Number

Knowing the number of live cells in a population permits one to determine the mutation frequency in terms of viable cells and to define the percentage of cells surviving a given treatment with mutagens or selective agents. Viability counts are useful in doing reconstruction experiments with putative mutant cell lines. With a cultured tobacco cell line, the percentage of dead cells in an untreated control population may be from 10 to 28% (unpublished observations from our laboratory). Widholm (1972d) noted that the viability of different species of plant cells in vitro dropped considerably when the growth rate decreased in the latter stages of culture.

Estimation of bacterial survival of treatments with mutagens or selective agents is accomplished by counting the number of colonies on agar plates. Variability in the plating efficiency of plant cells at low densities makes this a rather unreliable method for determining the number of viable cells in a population. Another method of determining cell survival is the extrapolation of growth curves of suspension cultures of plant cells back to the time of treatment (Sung, 1976). This method is particularly appropriate when the cultures contain highly aggregated cells.

Where the plant cells are highly dispersed, the percentage of live ones can be determined more rapidly by microscopic examination of cells that have been treated with a dye such as phenosafranine or fluorescein diacetate (Widholm, 1972d). The percentage of dead cells estimated by staining tobacco cells with 0.05% (w/v) phenosafranine correlated well with the percentage of dead cells determined by adding the percentage of cells with condensed cytoplasm to the percentage of cells that did not undergo plasmolysis (unpublished observations from our laboratory).

Research is in progress in various laboratories to determine effective mutagens for use with cultured plant cells. A significant proportion of spontaneous mutations that arise in bacteria are stable deletions (Drake, 1970). The selection of deletion mutations in plant cells may be desirable for some purposes. If deletion mutations in plant cells occur at a significant frequency in spontaneous mutants, their isolation may be feasible without the use of mutagens.

E. Use of a Liquid or Solid Medium for Selections

The selection of mutant plant cells can be carried out in a liquid or solid medium. Each type of medium has certain advantages and disadvantages. The cells that will undergo selection must be dispersed prior to the selection procedure. In some cases, callus that has not been in suspension culture is highly friable and falls apart into small cell clusters when it is placed into agitated liquid medium. In other instances, a suspension of small cell clusters or filaments can be derived from callus or tissue explants in liquid medium only after a prolonged period of subculture and selection of suspended cells. Whereas the callus from which such suspensions were derived may have been totipotent, the cells in suspension culture, like those of callus cultures (Narayanaswamy, 1977), have a tendency to lose their totipotency over time. The loss of totipotency may be accompanied by aneuploidy and polyploidization of chromosomes (Section IV,C). Another approach to the separation of cells is to sieve callus that has been recently initiated (Müller *et al.*, 1977). This may obviate the prolonged process of selecting a highly dispersed cell suspension and the concomitant loss of regenerative capacity.

Cells dispersed in an agitated liquid medium get a relatively equal exposure to the growth medium, whereas on an agar medium, chemical gradients develop. In addition to being more uniformly exposed to a mutagen or a selective agent, cells in suspension culture have a more rapid growth rate and presumably a higher mitotic index than cells on solid medium (e.g., see Polacco, 1976). Evaporation of the medium may not be as extensive over the shorter period of time and may be more readily controlled with liquid medium. Mutagen treatments and selections carried out in a liquid medium permit one to estimate the percentage of dead cell more easily (Section III,D) and to monitor any pH changes. A liquid medium permits the use of enrichment selection techniques (Sections V,A,C). Alone or in conjunction with agar-solidified medium, liquid medium allows the application of rescue techniques to isolate nongrowing mutant cells which survive the selective conditions that kill growing wild-type cells. Selections carried out in liquid may permit a quantitative determination of mutation frequency, provided that the initial cell number in each flask is considerably below the reciprocal of the estimated mutation frequency. There exists the possibility that two mutational events may occur in a single flask, however.

The use of agar plates for selecting mutant cell lines is advantageous in assuring the clonal nature of a mutant line and in quantitating the mutation frequency. In addition, the use of selective agar plates may aid in the early detection and removal of contaminants arising from fungal spores which may be troublesome during warm months.

The physiology of cells in a primary callus, in an established suspension culture, and in callus derived from the latter may be very different. A physiological difference between bacteria cultured on agar plates and bacteria cultured in liquid medium has been recognized. Hegeman (1966a) observed that on agar plates, but not in liquid medium, the growth substrate phenylacetate elicited the synthesis of enzymes that mediate the catabolism of mandelate in a fluorescent pseudomonad strain. Morphologically, plant callus cultures can be very different from the derivative suspension cell cultures. Biochemical differences are likely to be present also between plant cells cultured in liquid and those cultured on solid medium. One example of a gross nutritional difference is the requirement of plant cells for vitamins in liquid but not on solid medium (Torrey and Reinert, 1961).

IV. SOME ATTRIBUTES OF BACTERIA AND PLANT CELLS *IN VITRO*

A. Time Course of Growth

The growth curves of bacterial cells and cultured plant cells exhibit a lag period, an exponential phase, and a stationary phase of growth. The generation times of cells in culture are markedly different for bacterial and plant cells. The mean generation time of *E. coli* cells in liquid medium can be as short as 20 min. A cell line of *N. tabacum* var. Wisconsin 38, selected in our laboratory for its rapid growth rate in liquid medium, had a generation time of approximately 41 hr. Other rapidly dividing plant cell suspension cultures were reported to have mean doubling times of approximately 40 hr (Nishi and Sugano, 1970) and 48 hr (Filner, 1966; Jouanneau and Péaud-Lenoël, 1967). In liquid medium that selected against wild-type cells, a clone of mutant cells would not be recognized until it emerged from the background of the nongrowing wild-type population. Given the required minimum inoculation density (Section III,B) and a mean doubling time of 41 hr, it would take *at least* 24 days before the growth of a tobacco mutant cell line would be observed.

Cells taken at different phases in the growth curve differ metabolically. In addition, the total number of plant cells has a significant effect on the chemical composition of the medium due to the uptake and excretion of compounds by the cells. In devising a reproducible selection system for plant cells, whether using a liquid or solid medium, it is critical to control the variables of cell concentration and the cells' history as closely as possible.

B. Nature of Haploid Genomes

The recovery of recessive mutant strains requires that only one copy of the wild-type gene be present in the parental strain. In this section we briefly compare the bacterial and plant genomes with respect to this point. The reader is referred to recent reviews for the general differences between prokaryotic and eukaryotic DNA (Bonner, 1976; Key, 1976). The genome of a coliform bacterium or of a pseudomonad is generally assumed to be haploid. For some plant species, plantlets or callus having the gametic chromosome number can be obtained by anther or pollen culture (Sunderland, 1971, 1973b; Vasil, 1973; Nitsch, 1974; Smith, 1974) or by other methods used by plant breeders (Chase, 1969). Haploids can be diploidized in order to produce fertile plants (Burk *et al.*, 1972; Kasperbauer and Collins, 1972). The plant species *Lycopersicon esculentum* (tomato), *Zea mays* (corn), and two weeds, *Datura stramonium* and *Arabidopsis thaliana* are a few examples of plants that have long been the subjects of genetic studies; they have been identified as being true diploids. To our knowledge, the routine recovery of haploid callus or plantlets that can be continuously cultured *in vitro* has not been achieved with any of the above-mentioned species, although it has long been successful with *Datura innoxia* (Guha and Maheshwari, 1964, 1966, 1967) and other *Datura* species. Reports of success in producing haploid *Arabidopsis tha-*

liana (Gresshoff and Doy, 1972a), *Lycopersicon esculentum* (Sharp *et al.*, 1971, 1972; Gresshoff and Doy, 1972b; Debergh and Nitsch, 1973) and *Lycopersicon pimpinellifolium* (Debergh and Nitsch, 1973) *in vitro* suggests eventual achievement of consistent success in obtaining haploids of these species.

Many cultivated plant species are allopolyploids or autopolyploids (Allard, 1960). For example, the plant variety that has proved to be a model *in vitro* tissue culture system, *N. tabacum* var. Wisconsin 38, is an amphidiploid with a somatic chromosome number of 48. Anther culture of the variety yields an allodihaploid with 24 chromosomes. Goodspeed (1954) has postulated that the genus *Nicotiana* evolved from ancestral progenitors with six pairs of chromosomes.

Certain tissues of some plants are characterized by having cells with different levels of polyploidy (Bradley, 1954). Culture of explants from such tissues may give rise to cell lines that are highly polyploid. Shimada and Tabata (1967) showed this to be the case for the pith parenchyma of tobacco.

It is not possible to know *a priori* whether a given enzyme activity is encoded by more than one structural gene or whether alternative pathways for the metabolism of a compound exist. It may be that even "eudiploid" plant species contain nonallelic duplicates of certain structural genes or alternative metabolic pathways. These possibilities have been proposed as explanations for the failure to recover a wide spectrum of nutritional mutants in *Arabidopsis* and *Lycopersicon* (Li *et al.*, 1967). Conversely, it is possible that known amphidiploids are euhaploid for certain genes due to the loss of extra copies in the course of evolution. A clue might be obtained by examining different tissues of the plant and *in vitro* cultures for isozymic variants of the enzyme of interest. The presence of isozymic variants would suggest that multiple genes encode for the enzymatic activity, provided that proteolysis of the enzyme was ruled out.

C. Stability of Chromosomes

Discounting the instability of transmissible plasmids, the chromosome of a wild-type *E. coli* or *Pseudomonas* cell appears to be stable when the bacteria are cultured in the laboratory. In contrast, maintenance of plant cells *in vitro* is well known to be accompanied by changes in karyotype, including the generation of polyploidy, aneuploidy, endoreduplication, and structural aberrations (Dougall, 1972; Sunderland, 1973a; Sheridan, 1974; Narayanaswamy, 1977). In spite of this, cell lines that are stable with respect to chromosome number have been described (Reinert and

Kuster, 1966; Kao *et al.*, 1970; Palmer and Widholm, 1975), even after ten years of culture (Chu and Lark, 1976).

The factors that determine the karyotypic stability of cells *in vitro* remain elusive. It seems likely that any one of a number of parameters influences the behavior of chromosomes of cultured plant cells. A number of factors have been cited as contributing to the instability of chromosomes or the competitive advantage of cells with an altered chromosomal constitution. A few of these include subculturing plant cells after prolonged intervals of time rather than maintaining them in exponential phase (Bayliss, 1975), inclusion of agar in the medium (Halperin, 1966), or changes in growth substances in the medium (Melchers and Bergmann, 1958).

For some genetic studies, the generation of aneuploidy and chromosomal changes may be advantageous. However, endoreduplication and polyploidization of chromosomes undermines the selection of recessive mutations in haploid cells. Thus, monitoring the chromosome numbers of plant cells is important to the success of a selection for recessive mutants by indicating the proportion of cells that are haploid. In addition, the generation of karyotypic alterations during the prolonged culture of plant cells may be associated with a loss of morphogenetic potential (Murashige and Nakano, 1967; Narayanaswamy, 1977), although it does not necessarily preclude regeneration (Sacristán and Melchers, 1969).

D. Rare Phenotypic Alterations Unaccompanied by Changes in the Primary Structure of DNA

Phenotypic alterations may be mediated by fine controls involving the stimulation or inhibition of enzyme activity by allosteric effectors and by coarse controls involving regulation of enzyme biosynthesis at the level of transcription or translation. Included in transcriptional controls are induction, repression, and derepression of enzyme biosynthesis (Section V,A). In a bacterial population exposed to the same environment, the cells usually employ similar mechanisms of fine and coarse controls. In so doing, the population of cells adapts rapidly and relatively uniformly to changes in the external and internal milieu. A spontaneous point mutation or deletion mutation that permits a bacterial cell to adapt to the environment occurs in only one cell out of 10^5 to 10^9 cells. In order for a phenotypic alteration caused by the mutation to be expressed on the population level, many doublings of the mutant must occur. Thus, the adaptation mediated by a mutation requires hours for its expression at the bacterial population level, whereas the adaptation mediated by fine or coarse control is expressed on the population level within seconds or minutes, respectively.

Under some experimental conditions, a phenotypic alteration mediated by the coarse control of gene expression can mimic the occurrence of a mutation. A well-characterized example of this phenomenon in bacteria is provided by studies of the inducible galactoside permease and β-galactosidase in *E. coli* (Novick and Weiner, 1957). Induction of β-galactosidase was shown to be an all-or-nothing phenomenon. Under conditions of high inducer concentrations, the enzyme was distributed uniformly among individual cells in the population. Under conditions of low inducer concentration, only a small fraction of the cells in the bacterial population were induced for β-galactosidase synthesis, the majority remaining uninduced. The induced phenotype was a stable trait, being inherited by the progeny of the induced cells. The explanation for this phenomenon was that at low inducer concentration there was a random chance of the galactoside permease of a bacterium being induced. Once a single permease molecule appeared in a bacterium, the process of induction became autocatalytic (Novick and Weiner, 1957).

The regulatory controls that govern the differentiation of plant cells may be similar in mechanism to fine and coarse controls which mediate relatively rapid responses at the population level. The stability of the differentiated state, however, implies that the regulatory controls governing the cells of differentiated organisms are not altered readily. A type of stable phenotypic alteration may occur at a low frequency in cultured plant cells. The phenotypic change is heritable, but it is not believed to be mediated by a change in the primary sequence of DNA. The term "epigenetic" has been applied to such heritable but reversible changes in gene expression (Nanney, 1958); however, it should be noted that "epigenetic" is a catchall phrase for all types of regulatory control mechanisms that do not involve a change in the genetic sequence. Rare, heritable epigenetic alterations in phenotype are believed to arise at a frequency significantly higher than the frequency of spontaneous mutations (Gehring, 1968), and they are postulated to be less stable than mutations (Nanney, 1958). In cultured plant cells, the lower limit for the frequency of a rare epigenetic event was estimated to be 10^{-3} (Binns and Meins, 1973). The frequency of a heritable epigenetic event and reversion from it, however, may be as variable as mutation frequencies for different genetic loci. Thus, comparison of the frequency of occurrence is not a definitive index to the type of mechanism mediating phenotypic change. Moreover, the estimation of reversion frequencies depends on the system being amenable to the application of selection pressure for the reverse mutation or epigenetic event.

Totipotent plant cell cultures provide more definitive methods to distinguish between the two mechanisms. One method is to regenerate altered plant cells and reexamine callus or suspended cells derived from explants of the regenerated plants for the variant phenotypic trait. If the variation was caused by a mutation, it should continue to be expressed. If the trait was caused by an epigenetic event, the variant phenotype is likely to have reverted to the wild-type phenotype during the processes of differentiation and dedifferentiation. The most convincing analysis is made by subjecting regenerated variant plants to sexual crosses and screening the progeny plants for the variant trait.

A well-known example of epigenetic regulation occurring in cultured plant cells is the phenomenon of habituation. Habituation was originally described by Gautheret (1955), and the term generally refers to the heritable loss of a requirement for an exogenous supply of cytokinin or auxin. More recent studies using cloned cells showed definitively that habituation is inherited by individual cells *in vitro* (Lutz, 1971; Melchers, 1971; Binns and Meins, 1973). However, tissue from plants regenerated from cytokinin-habituated cells had lost the ability to grow *in vitro* in the absence of cytokinin (Binns and Meins, 1973). The early studies of Gautheret revealed that there are degrees of habituation. Binns and Meins (1973) found that clones of tobacco cells derived from a cytokinin-habituated tissue differed in their response to various levels of exogenous cytokinin. Optimal growth of the clones occurred at different concentrations of exogenous cytokinin. Thus, the epigenetic phenomenon of habituation appears to be characterized by a gradient of phenotypes rather than an all-or-nothing requirement for a growth regulator.

As mentioned in Section III,E, callus cultures of plant cells have been observed to differ morphologically and physiologically from derivative suspension cultures. Another type of nonmutant, phenotypic variant that may be observed in suspension cell cultures is a rare clone of cells with the phenotype of the callus or tissue from which the suspension was derived. The frequency of such variant cells in a population after extensive subculture may be as low as the mutation frequency. One way to ascertain whether this will pose a problem in selections is to maintain the original callus and subject it to the selective conditions that prevent growth of the derivative cells that have been in suspension culture. If the cells maintained as callus appear to be unaffected by the selective conditions, the possibility of recovering developmental variants must be recognized.

In conclusion, the distinction between mutant and nonmutant plant cell lines is not readily apparent because phenotypic alteration may not be caused by genetic mutation. It is becoming generally acknowledged that until a plant cell line having an altered phenotype has been convincingly demonstrated to be a *mutant cell line*, it should be described as a *variant cell line*.

E. Some Consequences of Genotypic and Phenotypic Heterogeneity of Plant Cell Suspension Cultures

The *in vitro* milieu does not reproduce the *in vivo* milieu of plant cells. It is likely that plant cells *in vitro* are subject to a continuous process of selection whereby the cell population is enriched with genetic mutants or epigenetic variants that are well adapted to the conditions imposed by the investigator. Genetic variation can be monitored in part by examining the karyotypes of cells *in vitro;* however, epigenetic variation is not as easy to monitor.

One consequence of the variation in cellular phenotype is that the growth rate of a plant cell line in a particular medium does not always remain a dependable constant. In addition, genetic and epigenetic heterogeneity in plant cell cultures may introduce complications into the study of the kinetics of biosynthesis of an enzyme in cells that have been transferred from a medium in which the enzyme is not expressed to a medium in which growth requires the activity of the enzyme. Where an increase in the specific activity of the enzyme occurs over a period of days rather than hours, the increase may be caused by the selection of cells that have more enzyme activity. For example, selection for epigenetic variants that are present at a high frequency could occur rapidly and mimic induction or derepression.

Because plant cells *in vitro* may contain unrecognized genetic defects as well as visually heterogeneous chromosomal complements, wild-type control cells should be regenerated as plants along with the variant cells. Regenerated wild-type control plants and the putative mutant plants can then be subjected to similar genetic and biochemical analyses to detect *in vitro*-induced changes that are not related to the effects of a putative mutation.

V. APPLICATION OF BACTERIAL SYSTEMS TO SOME PROBLEMS IN PLANT BIOLOGY

A. Regulation of Structural Gene Expression

The synthesis of bacterial enzymes occurs constitutively or is usually governed by one of two types of regulation: repression and derepression or induction and catabolite repression. The rate of synthesis of an enzyme that is produced constitutively remains constant relative to that of other proteins under various conditions of growth. Repression and derepression control the synthesis of many of the enzymes of biosynthetic pathways in bacteria and permit the rate of synthesis of an enzyme to vary. In the enteric bacteria, for example, the synthesis of the tryptophan biosynthetic pathway enzymes varies over 100-fold by means of repression and derepression (Crawford, 1975). It is not known to what extent the biosynthesis of plant enzymes is regulated by repression and derepression.

An enzyme is termed inducible if its rate of biosynthesis is increased relative to general protein synthesis in the presence of certain compounds. In the case of plants, an environmental stimulus such as light may serve to initiate an increase in specific *de novo* protein synthesis. An inducer is the growth substrate, metabolite, metabolic analogue, or environmental factor that most directly elicits the increase in the rate of synthesis of an enzyme. Many bacteria can utilize a range of growth substrates, and most of the catabolic enzymes that mediate the dissimilation of the compounds are inducible (Jacob and Monod, 1961; Ornston, 1971). The rates of biosynthesis of the enzymes may be increased from 30-fold to more than 1000-fold in the presence of their inducers.

There is evidence that a number of plant proteins that are amenable to study in cell suspension culture are inducible. Some of these are included in the review of Filner *et al.* (1969). In addition, phenylalanine ammonia-lyase appears to be an inducible enzyme (Creasy and Zucker, 1974; Hahlbrock and Schröder, 1975). Other enzymes of phenylpropanoid synthesis and those of flavonoid and lignin synthesis (Grisebach and Hahlbrock, 1974) may prove to be inducible. Several enzymes and an uptake system involved in the assimilation of nitrogen appear to be inducible in the species of higher plants studied. These include urease (Matsumoto *et al.*, 1966; Bollard *et al.*, 1968; Polacco, 1976), nitrate reductase (Hewitt *et al.*, 1976), nitrite reductase (Kelker and Filner, 1971; Stewart, 1972), and the nitrate uptake system (Heimer and Filner, 1970, 1971; Jackson and Yanofsky, 1973). To date, a few of the putatively inducible enzymes of plants appear to be substrate induced; some of these have been shown to be induced by hormones, light, or certain levels of sucrose as well. In this review we will be concerned with induction by substrate or product, as defined below (Section V,A,2).

The mechanisms of induction and the distinction between true induction and apparent induction are summarized in Section V,A,1. With this background we limit our discussion to the identification of the inducer of an enzyme, using bacterial systems as models. The extent of inductive control will not be covered; the reader is referred to a recent review treating this topic as it applies to bacteria (Ornston and Parke, 1977). The section will be concluded by a discussion of the identification of inducers in plant cells by means of a physiological genetic approach.

1. Mechanism of Induction of Enzyme Synthesis

A number of mechanisms could account for an increase in the specific activity of an enzyme in response to a particular compound or stimulus. Studies with bacteria have revealed two types of regulatory control: positive or negative control of genetic transcription. One example of negative control of genetic transcription is provided by the lactose operon of *E. coli* (Jacob and Monod, 1961). A repressor protein binding to the operator gene prevents the transcription of the structural genes for the operon in the absence of inducer. The interaction of the inducer with the repressor protein triggers transcription because the repressor protein no longer binds to the operator site. An example of positive control of genetic transcription is provided by the L-arabinose operon, *ara*BAD, in *E. coli* (Englesberg and Wilcox, 1974). A regulatory gene product, an activator protein, must interact with the initiator site in order for transcription of the operon to occur. Extensive genetic criteria have been applied to establish negative control of the lactose operon (Jacob and Monod, 1961) and positive control of the *ara*BAD operon (Englesberg and Wilcox, 1974). Transcription and translation appear to be temporally (Jacob and Monod, 1961) and spatially (Hamkalo and Miller, 1973) coupled in bacteria. The primary event in the induction of enzyme synthesis in bacteria is believed to occur at the level of transcription.

The regulation of inducible enzyme synthesis is likely to be mediated by different mechanisms in higher plant cells in contrast to bacteria. Synthesis of messenger RNA, which occurs in the nucleus, is temporally and spatially separated from the synthesis of protein in the cytoplasm. The presence of stable messenger RNA in higher organisms raises the possibility of an inductive event occurring at a posttranscriptional level.

An increase in the specific activity of an enzyme in the presence of a putative inducer could be due to the activation of macromolecular precursors or a decline in the rate of degradation of an enzyme rather than to *de novo* protein synthesis. The activation could be mediated by proteolysis (Northrup *et al.*, 1948); by the interaction of the enzyme's substrate, a modulating metabolite, or a trace element with a protein precursor (Cazzulo *et al.*, 1969; Rigano, 1971; Vega *et al.*, 1971); or by a decrease in the amount of an enzyme inhibitor (Pressey and Shaw, 1966). Filner *et al.* (1969) have pointed out that a decrease in the rate of degradation of an enzyme may mimic induction. The substrate of a preexisting enzyme may protect it from degradation (Schimke *et al.*, 1965). Such a mechanism of apparent substrate induction may be particularly important in proliferating higher plant cells where the rate of protein turnover is high (Zielke and Filner, 1971) as opposed to rapidly growing bacterial cells where the rate of protein turnover is low (Schimke and Doyle, 1970). Inhibitors of protein or RNA synthesis have been used to study whether an increase in the specific activity of an enzyme is mediated by *de novo* protein synthesis or activation of a protein precursor (e.g., Afridi and Hewitt, 1965; Ingle *et al.*, 1966). However, the inhibition of specific enzyme synthesis by these agents is not proof of *de novo* protein synthesis since they might act indirectly. For example, they could interfere with the synthesis of a protease.

A convincing demonstration of *de novo* enzyme biosynthesis in higher plants requires physiochemical evidence. Radioisotope labeling and purification of an enzyme or density labeling and isopycnic equilibrium centrifugation can be applied to this end (Filner *et al.*, 1969). In addition, the translation of isolated polyribosomal mRNA in a heterologous cell-free system has been used to study the mechanism of synthesis of plant protein; antiserum prepared against the purified protein is used to monitor the specific mRNA. Evidence collected by this technique suggested that the induction of phenylalanine ammonia-lyase in parsley cell cultures was mediated by an increase in the concentration of translatable mRNA specific for the enzyme (Schröder *et al.*, 1977). Working with tobacco suspension cultures, Zielke and Filner (1971) used density labeling and isopycnic equilibrium centrifugation to establish that nitrate reductase activity appears in response to nitrate as a result of *de novo* protein synthesis. However, the density labeling technique did not permit them to rule out that the specific activity of nitrate reductase increases in the presence of nitrate because of a decrease in the rate of degradation of the enzyme.

2. Identification of Inducers in Bacteria

Basic to an understanding of the regulation of the biosynthesis of an inducible enzyme is the identification of the inducer. Prior to identifying an inducer, it is essential to understand the biochemistry of the metabolic sequence being studied. In dissimilatory pathways, the growth-supporting activity of enzymes moves from growth substrate to common intermediary metabolites. This direction defines the substrates and products of the enzymes. A given enzyme in a metabolic sequence of reactions may be subject to *substrate induction* by the substrate of the first enzyme in the sequence or it may be subject to *product induction* by the product of the last reaction in the sequence. All product inducers also serve as substrate inducers in bacteria (Ornston and Parke, 1977).

The definitive identification of an inducer requires that the putative inducer be nonmetabolizable. Three methods of achieving this goal have been employed with bacteria. (a) The natural compound may be chemically altered to an inducer analogue that does not serve as a substrate for an enzyme. (b) In some cases, a physiological restriction that prevents the metabolism of a putative inducer may be imposed. This is particularly useful when it is not possible to block the metabolism of a compound by genetic means. (c) A third method is the genetic modification of the cells so that they cannot metabolize the putative inducer. Elucidation of the regulation of enzymes may be complicated by permeability barriers. Some inducers and inducer analogues may not effect enzyme induction because the cells are insufficiently permeable to the compounds. In such cases, it may be necessary to select permeable mutant strains.

a. Nonmetabolizable Inducer Analogues. The use of inducer analogues is illustrated by the identification of L-kynurenine as the inducer of the first three enzymes in the L-tryptophan dissimilatory pathway in a fluorescent pseudomonad. Palleroni and Stanier (1964) reduced the γ-ketone of L-kynurenine to a hydroxyl group. The compound could not be hydrolyzed by kynureninase to pathway intermediates, but it served as an effective substrate and product inducer of the first three enzymes of the pathway.

The identification of the true inducer of the *lac* operon in *E. coli* was aided by the use of the inducer analogue, isopropyl-β-D-thiogalactopyranoside (IPTG). Comparative studies of the kinetics of induction by IPTG and by the putative inducer, lactose, suggested that a metabolic transformation was required to convert lactose into an inducer. Subsequent research revealed that allolactose was the natural inducer (Jobe and Bourgeois, 1972).

b. Imposition of a Physiological Restriction. A physiologically restrictive condition was employed by Kemp and Hegeman (1968) to determine whether catechol was an inducer of the enzymes of the β-ketoadipate pathway in *Pseudomonas aeruginosa*. The cleavage of the aromatic ring of catechol is mediated by catechol oxygenase in the presence of its two substrates, catechol and oxygen. The capacity of *P. aeruginosa* cells to respire nitrate was exploited. When cells were grown in the presence of catechol with lactate as the growth substrate and nitrate as the sole terminal electron acceptor, catechol could not be metabolized. Because the enzymes of the β-ketoadipate pathway were not induced, Kemp and Hegeman could conclude that catechol did not serve as an inducer of them.

c. Mutant Strains That Cannot Metabolize a Putative Inducer. The third method of rendering a putative inducer nonmetabolizable is to select a mutant strain that lacks an enzyme for which the inducer is a substrate. A great number of selective systems have been employed to accomplish this goal. A direct method of selecting mutant strains of bacteria is to plate mutagenized cells onto a medium containing very low levels of a nonselective growth substrate and high levels of a selective growth substrate (Palleroni and Stanier, 1964). Mutant colonies that cannot grow at the expense of the selective growth substrate appear as small colonies. Subsequent plating of cells from these colonies onto nonselective plates and replica plating them onto selective plates by the method of Lederberg and Lederberg (1952) identifies those colonies that are unable to metabolize the selective growth substrate. Another direct method of selection involves the use of an inducer analogue that inhibits the normal induction of an enzyme. The selection of cells that are resistant to an antiinducer may lead to the recovery of negative regulatory mutants that can no longer synthesize the enzyme that acts on the putative inducer. Other mutants may be selected by this method as well, however, for instance, cells with altered inducer specificity (Cánovas et al., 1968).

An effective method of indirectly selecting spontaneous mutant strains dysfunctional in a biosynthetic or catabolic enzyme activity employs penicillin treatment and replica plating. Penicillin is an agent that interferes with the synthesis of the peptidoglycan cell wall of prokaryotes. Growing, wild-type bacterial cells lyse in a medium containing penicillin (Davis, 1948; Lederberg and Zinder, 1948) or, more effectively, the two cell wall agents penicillin and D-cycloserine (Ornston et al., 1969). The repeated transfer between the penicillin medium and an alternative nonselective growth medium results in an *enrichment* of mutant cells that do not grow in the penicillin medium. The spontaneous mutant cells may account for over 1% of the mixed population (Ornston et al., 1969). Subsequent plating of the mixed population of cells onto nonselective plates and replica plating the colonies onto plates containing selective media differentiates the mutant and wild-type colonies.

In the direct and indirect selection schemes, a primary growth substrate or a metabolite formed in its catabolism can be used in screening and analyzing mutant strains. Potential difficulties may arise if a metabolite is chemically unstable in the growth medium. If a metabolite does not permeate the cell membrane readily, it may be necessary to select permeability mutant strains. Genetically blocking the metabolism of a putative inducer may be complicated (a) if the synthesis of the compound is mediated by a physiologically reversible reaction; (b) if more than one structural gene encodes for an enzyme activity that catabolizes the inducer; or (c) if the putative inducer serves as a substrate for more than one enzyme activity.

3. Identification of Inducers in Plant Cells

Physiological genetic studies of cultured plant cells are complementary to work with whole plants or differentiated tissues. Identification of putatively inducible enzymes has been achieved largely with differentiated plant tissues (e.g., see Filner *et al.,* 1969). The use of cultured plant cells permits careful control over the environment and the history of the cells. It has permitted the study of the kinetics and extent of induction under conditions where the nitrogen source and concentration are defined and relatively uniform for each cell in the population. The increase in specific activity as a result of the transfer of cells from a noninducing to an inducing medium was 1500-fold for nitrate reductase of cultured tobacco cells (Filner, 1966) and 100-fold for urease of cultured soybean cells (Polacco, 1976).

The use of cultured plant cells offers the possibility of selecting for mutant cells that cannot metabolize a putative inducer. The selection of mutant plant cell lines blocked in the assimilation of nitrate (Müller *et al.,* 1977) will be useful supplements to the use of tungstate (Kelker and Filner, 1971; Heimer *et al.,* 1969; Wray and Filner, 1970) as a physiological probe to identify the true inducers of the nitrate assimilatory uptake system and enzymes. Tungstate renders nitrate reductase inactive, presumably because it prevents the incorporation of molybdate into the nitrate reductase enzyme complex.

Rendering the inducers in other metabolic pathways nonmetabolizable may require the selection of mutant strains. A lack of suitable analogues or of physiologically restrictive conditions may necessitate this. There are several potential difficulties in selecting mutant cell lines that are dysfunctional for a particular enzymatic activity. Agents analogous to penicillin or D-cycloserine that interfere with the biosynthesis of plant cell walls have not yet been identified. However, alternative negative selection schemes have been developed. 5-Bromodeoxyuridine (BUdR) has been used to select for auxotrophic mutant cell lines of tobacco (Carlson, 1970). Incorporation of BUdR into the DNA of cells causes the cells to be inactivated by visible light (Puck and Kao, 1967). Chu and Lark (1976) found that 5-fluorodeoxyuridine could be used to synchronize mitoses in soybean suspension culture cells. Exposure of the synchronized cells to BUdR and subsequently to light resulted in the death of cells synthesizing DNA. The use of arsenate as a counterselective agent to kill replicating cells is being investigated (J. C. Polacco, personal communication). The employment of single counterselective agents raises the possibility of unintentionally selecting mutant cells that are resistant to them. In fact, mutant plant cells resistant to BUdR have been selected (Ohyama, 1974;

Márton and Maliga, 1975). The use of more than one agent simultaneously would decrease the chances of selecting resistant cells.

The counterselection of wild-type cells may be complicated by the presence of nondividing wild-type cells in the population. Chu and Lark (1976) observed that 20 to 30% of the cells in a suspension culture of soybean cells were capable of synthesizing protein and RNA, but they did not synthesize DNA or divide. The cells that were alive but not dividing could recover the ability to divide. If the presence of nondividing wild-type cells in suspension cultures is a universal phenomenon, the ratio of mutant cells to viable wild-type cells present after a round of counterselection may be lowered considerably.

A reliable replica plating method for plant cells does not exist (however, see Schulte and Zenk, 1977). In addition, spot tests that do not kill plant cells and that are effective in the face of diffusion remain to be developed. The absence of these two techniques makes the discrimination of mutant from wild-type cells a difficult procedure with plant cells.

The selection of plant cell mutants that are totally negative for an enzymatic activity may be hampered because of the amphiploid nature of the genome of some plant species. Ideally, genetic selections should be performed with a haploid cell line of a plant species that has been identified as a true diploid by classical breeding techniques. In practice, cultured cells of the amphiploid *N. tabacum* have been used in a number of genetic studies. The ease of obtaining haploid cell lines, culturing them, and regenerating whole plants has made this a model system in all respects but genetic. All of the auxotrophic cell lines of the allodihaploid *N. tabacum* isolated by Carlson (1970) were leaky. On the other hand, Müller *et al.* (1977), working with an allodihaploid cell line of *N. tabacum,* were able to select mutant cell lines that were totally lacking nitrate reductase activity. Out of a total of 62 stable chlorate-resistant cell lines, 15 clones were dysfunctional for nitrate reductase activity, while the others were leaky. It will be of interest to learn whether the cell lines that have no nitrate reductase activity possess dominant or recessive mutations and whether the mutations are structural or regulatory. In fungi, a dominant regulatory mutation resulted in the inability of cells to synthesize an enzyme (Mahadevan and Eberhart, 1962).

In order to select mutant cell lines, it is usually necessary to find stringent conditions that restrict the growth of wild-type cells or, in the case of counterselection, mutant cells. In the latter case, it is also essential to have nonrestrictive growth conditions. Study of the inducible nature of enzymes involved in the assimilation of primary nitrogen sources in plant cells is feasible, since a number of different nitrogenous compounds can serve as the sole nitrogen source for plant cells *in vitro*. It may be more difficult to investigate the regulation of enzymes that participate in the catabolism of carbon compounds, since plant cells generally utilize only a few related carbon compounds as the sole source of carbon and energy. In studying the metabolism of sucrose and related carbon sources, temperature rather than substrate could be the basis of restrictive or nonrestrictive conditions. The technique of selecting temperature-sensitive mutations in cultured plant cells has been discussed elsewhere (e.g., Chu and Lark, 1976).

Plant cells may require unique solutions to the problem of recovering mutant clones that are blocked in the catabolism of metabolites that do not serve as primary growth substrates. One solution would take advantage of the fact that plant cells do not grow with ammonia as the sole nitrogen source unless succinate or an intermediate of the tricarboxylic acid cycle is present in the growth medium (Section III,A). Provision of a compound that is catabolized to an intermediate of the tricarboxylic acid cycle would enable plant cells to grow with ammonia as a sole nitrogen source and would permit the selection of mutant cells that cannot metabolize the compound.

B. Elucidation of Metabolic Pathways and Uptake Systems

The physiological and genetic techniques used to render an inducer nonmetabolizable (Section V,A) apply equally well to the establishment of the step reactions of a metabolic pathway and the properties of an uptake system. The step reactions can be initially investigated by identifying the intermediates of the pathway. Mutant strains that lack an enzyme activity frequently accumulate an intermediate or intermediates of the blocked pathway. In addition, the response of auxotrophic mutants to various putative intermediates in a blocked biosynthetic pathway may indicate the nature of the compounds involved in the pathway. The use of such strains has aided greatly in the elucidation of metabolic pathways in microorganisms (Greenberg, 1969; Rodwell, 1969).

Effective characterization of an uptake system requires that the substrate of the uptake system be rendered nonmetabolizable. This allows measurement of the intracellular concentration of substrate without the complications arising from its further metabolism (Kepes and Cohen, 1962). Uptake systems are studied by using nonmetabolizable substrate analogues, a physiological block of substrate metabolism, or mutant strains that are dysfunctional for the enzyme that initially attacks the substrate.

Similar approaches to the study of metabolic pathways and uptake systems that are expressed in cultured plant cells may be feasible. As mentioned above (Section V,A,3), the use of tungstate to block the metabolism of nitrate aided in a study of the nitrate uptake system in cultured plant cells (Heimer *et al.,* 1969; Heimer and Filner, 1971). However, most studies of uptake systems in higher plants may be complicated by the metabolism of the substrate. The potentials and pitfalls of selecting mutant plant cells that do not metabolize a particular substrate have been discussed in the previous section (Section V,A,3).

C. Genetic Modification of the Level of Production of a Structural Gene Product

Mutations that increase the rate or level of synthesis of a structural gene product are extremely valuable. They assist in the purification or characterization of structural gene products by causing the initial specific activity to be higher than it would be in wild-type cells. Another reason for selecting hyperproducers is to enhance the growth rate of organisms that may be limited by the quantity of the system of interest. Mutations to hyperproduction of an enzyme may increase the level of a product of enzymatic activity. Creation of a duplicated gene may provide material for further evolution. As discussed by Carlson and Polacco (1975), the preferential increase in the production of a plant enzyme or protein that is rich in a nutritionally limiting amino acid may lead to the recovery of a whole plant that reflects the increase in its edible structures.

Increases in specific proteins in microbes can be generated by the following genetic lesions: an increase in the number of gene copies, mutation at a regulator gene, mutation at an operator locus, mutation at a promotor site of the operon, or a lack of specific degradation systems. Spontaneous translocation–duplication mutations that create copies of a structural gene under separate transcriptional control can occur with the frequency of point mutations in bacteria (Jackson and Yanofsky, 1973) and yeast (Hansche, 1975). Methods of selecting for the different mutations in bacteria include (1) selection of mutants in a chemostat, (2) selection of mutants that produce a protein constitutively rather than inducibly, and (3) selection of cells resistant to an end-product repressor. These three topics will be discussed in more detail below. In addition, production of an inducible enzyme may be elevated if the metabolism of its inducer is blocked in mutant strains (Section V,A) or if the inducer is synthesized in elevated amounts in feedback-resistant mutant strains (Gross, 1969). The reader is referred to Demain (1971) for the topic of a plasmid- or phage-mediated increase in the number of gene copies in bacteria.

1. Selection of Mutants in a Chemostat

Gene duplication mutations have been observed to arise in bacteria (Horiuchi *et al.*, 1962; Rigby *et al.*, 1974) and yeast (Hansche, 1975) that have been grown in a chemostat on limiting concentrations of growth substrate. Such mutant strains possess a gene duplication that causes an increase in the production of a protein with an essential activity. Under selective pressure in a chemostat, mutant strains that have a small increase in growth rate may supplant the parental strain (Powell, 1958).

The selection of mutant plant cells in a chemostat may be feasible, but some potential problems should be addressed. The exclusion of contaminants during the lengthy time course for mutant plant cell selection in a chemostat is likely to be difficult. Ideally, the parental plant cell line should grow as a homogeneous suspension of single cells. The inadvertant selection of cells that clump or adhere to the culture apparatus might interfere with the selection of desired mutant cells. Mutations affecting cell clumping were encountered in selecting mutant strains of yeast in a chemostat (Francis and Hansche, 1972, 1973). The capacity of plant cells to undergo normal morphogenesis may be lost during their prolonged culture in a chemostat. The identification of mutant cells in the population requires an effective cloning and screening procedure to differentiate mutant from wild-type cells. The release of metabolites from viable and dead cells into the medium may raise the concentration of the limiting nutrient. Due to the relatively high ratio of total cellular volume to the volume of culture medium, it may be necessary to starve the plant cells partially for the nutrient that will be limiting in the chemostat.

In spite of the anticipated difficulties of selecting mutant plant cells in a chemostat, it appears likely that gene duplication mutations will not revert as frequently as they do in bacteria. Many gene duplications in bacteria are unstable, reverting much more rapidly than they are formed (Hill *et al.*, 1969; Jackson and Yanofsky, 1973; Rigby *et al.*, 1974). Rigorous selective conditions are required to maintain them in a population. Although a spontaneous gene duplication selected for in a chemostat was observed to occur at a low frequency of about 10^{-11} in a haploid yeast, the duplication was much more stable than bacterial gene duplications similarly selected (Hansche, 1975). Evidence indicated that the duplicated yeast structural gene had been transposed to a linkage group that segregated independently from the original gene. Results with the eukaryote yeast may not be directly applicable to higher plant cells. However, there is evidence that the haploid genome of a plant possesses stable duplicated structural genes (Schwartz, 1966) and that plant cells contain repetitious DNA sequences to a greater extent than bacteria (Britten and Kohne, 1968; Bonner, 1976; Key, 1976).

2. Selection of Constitutive Mutant Strains

A mutation leading to the constitutive synthesis of an inducible enzyme or uptake system can sometimes lead to the production of large quantities of the protein under some growth conditions (Hegeman, 1966b; Clarke, 1974; Parke and Ornston, 1976). Growth in a chemostat on limiting levels of a substrate inducer can lead to the selection for constitutive mutants that produce elevated levels of an enzyme (Novick and Horiuchi, 1961; Hegeman, 1966b). As discussed by Demain (1971), the use of a poorly inducing substrate or an inhibitor of induction can also select for constitutive mutants that may overproduce an enzyme.

Another effective method for the isolation of constitutive mutant strains of bacteria is the sequential transfer of cells between media containing a noninducing growth substrate and media containing an inducing growth substrate as a sole carbon and energy source. After repeated transfers, the culture may become enriched for constitutive mutant strains. This procedure was used first by Cohen-Bazire and Jolit (1953) to obtain mutant strains of *E. coli* that produced enzymes of the *lac* operon constitutively. Subsequently, Parke and Ornston (1976) used an analogous method to obtain *Pseudomonas putida* strains that formed constitutively enzymes and an uptake system that had been inducible by β-ketoadipate. The unique regulatory control of the β-ketoadipate pathway revealed that the selective pressure for constitutivity operated at the level of an uptake system for the inducing substrate, β-ketoadipate. Constitutive synthesis of the β-ketoadipate uptake system was accompanied by its hyperproduction.

A number of differences between bacterial and plant cells should be considered before applying the sequential transfer of plant cells to the selection of constitutive mutant cells. As discussed in Section III,B, the number of cell generations that can be supported in each flask is considerably greater for bacteria than for plant cells. Thus enrichment for mutant plant cells would involve many more transfers and more flasks. The level of the inducing compound supplied may be critical to the success of the sequential transfer method. Extremely high levels of an inducing carbon source or nitrogen source, as are used in numerous plant media, may not be as effective as lower levels. In addition, it may be important that cells separate after dividing.

Chemostat culture and the sequential transfer of cells are enrichment techniques. They require that cells be cloned and screened to differentiate mutant from wild-type cells. Bacterial selections involving chemostat culture or the sequential transfer of cells between inducing and noninducing media have been used most frequently to select for mutations specific for catabolic functions. It should be possible to expand their application to the selection of mutations in the assimilation of nitrogenous compounds or other nutrients. The nature of induction of enzymes in plant cells may differ from that in bacteria (Section V,A,1). Consequently, mutant plant cells selected by the sequential transfer method may have different characteristics from those of bacterial mutants similarly selected, for example, a loss of a specific enzyme degradation system. With ingenious modifications, the sequential transfer technique may be applicable to suspension cultures of plant cells.

3. Selection of Mutant Strains Resistant to Repression

Cells resistant to catabolite or end-product repression of an enzyme may have higher levels of the enzyme (Clarke and Lilly, 1969; Demain, 1971). One method of selecting for mutants resistant to catabolite repression is to cycle cells between a medium containing a catabolite repressor as substrate and a medium containing a substrate whose metabolism is subject to catabolite repression (Demain, 1971). The principles behind the sequential transfer of cells between two media discussed above (Section V,C,2) apply to this method of selecting cells insensitive to catabolite repression. Use of a toxic antimetabolite to select strains resistant to end-product repression (Umbarger, 1971) can be readily applied to those plant enzymes that are expressed *in vitro* and that are repressible. Putative mutant clones are identified as those that grow on the selective medium, thus avoiding the need for spot tests on mechanically cloned cells.

This section has touched on only some of the selective techniques that may be used to isolate plant cells with an elevated production of a structural gene product. The factors considered in applying bacterial selections to cultured plant cells should be relevant to other selective systems of interest. It is likely that some classes of mutations selected by the above methods would be dominant or semidominant. The use of a euhaploid plant cell line, however, would ensure the recovery of recessive mutations as well.

D. Alteration of the Primary Sequence of a Structural Gene Product

1. Selection for Structural Gene Mutations in Bacteria

In addition to causing a loss of catalytic activity, structural gene mutations can result in an alteration in the K_I of an enzyme for a particular inhibitor or a change in the K_m or V_{max} of an enzyme for a given substrate. The selection of mutants having an enzyme with an altered K_I has focused largely on biosynthetic pathways. Regulation of the biosynthesis of most amino acids in bacteria occurs by repression and derepression and by feedback inhibition. Commonly, the first reaction in the pathway is subject to inhibition by the end-product amino acid. The use of amino acid analogues as selective agents has permitted the isolation of amino acid analogue-resistant strains. One class of amino acid analogue-resistant mutants has an enzyme that is insensitive to feedback inhibition by the end product (Umbarger, 1969, 1971); the loss of end-product control may lead to an increase in the pool size of the amino acid. Umbarger (1971) has pointed out that the same analogue may select for different types of mutant phenotypes in different species. For example, in one species, a given analogue might select predominantly for an end-product-insensitive enzyme, while in a different species, the same analogue would select for an operon that was derepressed. Thus, the use of appropriate analogues appears to be critical in the selection of mutant strains having a regulatory enzyme with an altered K_I.

In addition to affecting the K_I, mutations that alter the primary sequence of an enzyme may affect the affinity of the enzyme for a substrate (K_m) or the rate of catalysis (V_{max}). Selecting for growth of *Klebsiella aerogenes* on low levels of xylitol in a chemostat, Hartley and his associates found that duplication of the ribitol dehydrogenase gene permitted the organisms to grow slowly at the expense of xylitol (Rigby *et al.*, 1974). However, screening of a total of 10^{14} bacteria failed to isolate strains with altered enzyme kinetics for xylitol. Subsequently, they discovered that mutagenesis with high levels of nitrosoguanidine produced multiple mutations in ribitol dehydrogenase leading to an enzyme with an altered K_m (apparent) for xylitol (Hartley, 1974).

The aliphatic amidase of *P. aeruginosa* provides another example of an alteration in enzyme properties as a result of selective pressures (Clarke, 1974). Acetamide and propionamide serve as good inducers and substrates of the enzyme in wild-type strains. Following the selection of a regulatory mutant strain that produced amidase under the selective conditions employed, Clarke and her colleagues isolated mutant strains capable of cleaving other aliphatic amides. Some of these mutant strains possessed an amidase with an altered K_m or V_{max} (Brown *et al.*, 1969). Brown and Clarke (1972) prepared chymotryptic and tryptic peptide maps of the wild-type and a mutant amidase and showed that the altered amidase contained a single amino acid substitution. The selection of mutant amidases that acted on some other amides required more than a single structural gene mutation in addition to an initial regulatory gene mutation (Brown *et al.*, 1969; Clarke, 1974).

Most enzymes are not totally specific for a single substrate; they may possess a low level of activity toward other similar compounds. The ability to utilize a new compound is frequently acquired by bacteria as a result of a mutation in a regulatory gene that governs the biosynthesis of an enzyme or a transport system that acts on the compound. Since most of the peripheral catabolic enzymes of bacteria are induced by a specific growth substrate or metabolite, expression of the enzyme in the presence of a new substrate or its metabolites may require a mutation to constitutivity or to a change in the specificity of induction. Subsequently, "fitter" mutations may increase the rate of catalysis and/or the affinity of the protein for the new substrate. Under similar selective pressures, for example, growth on a limiting carbon source in a chemostat, different physiological units of function respond in diverse ways. The aliphatic amidase of *P. aeruginosa* studied by Clarke often acquired new catalytic activities as a result of one or two mutations in its structural gene. On the other hand, the ribitol dehydrogenase investigated by Hartley acted upon "unnatural" pentitols at an increased rate in a *K. aerogenes* cell due to the duplication of its structural gene, and changes in catalytic specificity of the enzyme required multiple mutations in its structural gene.

2. Selection for Structural Gene Mutations in Cultured Plant Cells

Advances made in the selection of bacterial enzymes with altered sequences should be instructive to the plant somatic cell geneticist. Amino acid analogue-resistant plant cell mutants have value in elucidating the regulation of amino acid metabolism and, like other mutations, in providing genetic markers. Plants are similar to bacteria in controlling the biosynthesis of many amino acids by means of feedback inhibition of the first enzyme in the pathway (Miflin, 1973). The selection of mutant plant cell lines that have increased pools of certain amino acids due to a loss of feedback control has been widely recognized as being a step toward the recovery of crop plants that are enriched for nutritionally limiting amino acids.

The selection of amino acid analogue-resistant strains of microorganisms has provided models for the selection of plant cell lines resistant to amino acid analogues (Widholm, 1972a,b,c, 1974b, 1976, 1977; Carlson, 1973; Chaleff and Carlson, 1974b; Palmer and Widholm, 1975). Many of the variant cell lines were characterized as having increased levels of particular free amino acids. Of the cell lines that exhibited increased pool sizes, three methionine sulfoximine-resistant tobacco cell lines (Carlson, 1973) and a 5-methyltryptophan-resistant carrot cell line (Widholm, 1974a,b) were regenerated into whole plants. The trait appeared to be semidominant in plants regenerated from two of the methionine sulfoximine-resistant cell lines, and young leaves from the plants had five times the wild-type concentration of free methionine. Biochemical characterization of the regenerated mutant carrot cell line revealed that 5-methyltryptophan resistance was correlated with an impairment in the uptake of 5-methyltryptophan (Widholm, 1974a,b). Unexpectedly, the uptake mutant had an increased level of free tryptophan. Tobacco and carrot cell lines resistant to p-fluorophenylalanine also exhibited a decreased uptake of p-fluorophenylalanine and phenylalanine, which was accompanied by a greater phenylalanine pool in the carrot cells (Widholm, 1974b; Palmer and Widholm, 1975). It may be that the pool sizes were higher in the cell lines with impaired uptake because the rate of efflux of the amino acids was reduced as well. It should be mentioned that uptake mutants need not possess structural gene mutations.

Biochemical characterization of other amino acid analogue-resistant cell lines has been carried out by Widholm. Since the selections were conducted with diploid cell lines, putative mutations are likely to be present in a heterozygous condition. The resistant cell lines appear to be true mutants because of the low frequency of spontaneously arising resistant cells (6×10^{-5} or lower) and the stability of the variant trait in a nonselective medium (Widholm, 1974b, 1977). Regeneration of the cell lines to verify this was not possible because the cells had been maintained *in vitro* for a long period of time and had lost their totipotency (Widholm, 1974b). The biochemical studies of Widholm have identified anthranilate synthetase, the first enzyme of the tryptophan biosynthetic pathway, to be the site of the phenotypic alteration in many 5-methyltryptophan-resistant cell lines. The K_I of the enzyme with respect to the end-product inhibitor was altered; the change in K_I was accompanied by an increase in the level of the free amino acid. Variations in the K_I of anthranilate synthetase in tobacco and carrot cell lines resulted in increases in the levels of free tryptophan of 33- and 27-fold, respectively, over the wild-type cell lines (Widholm, 1972b,c). Cells resistant to lysine, methionine, and proline analogues showed elevated levels of the corresponding amino acids (Zenk, 1974; Widholm, 1976, 1977), but the biochemical basis of the increased pool sizes has not yet been established.

Our treatment of this promising area of plant cell research has been only cursory, since several reviews have discussed the selection of amino acid analogue-resistant plant cell lines clearly and concisely (Chaleff and Carlson, 1974a; Widholm, 1974b, 1977). It should be noted that genetic studies with the eukaryote yeast have revealed that all mutations rendering an amino acid biosynthetic enzyme insensitive to feedback inhibition are dominant (Bussey, 1970; Rasse-Messenguy and Fink, 1973). One would expect the selection of feedback-insensitive mutant plant cell lines to be facilitated by the use of diploid cells, since recessive analogue-resistant mutants would not be selected for.

Mutations modifying the catalytic specificity, K_M, or V_{max} of a plant enzyme would be useful for a number of reasons. The detoxication of specific herbicides by a plant could be achieved by the evolution of enzymes that act on them. Alterations in the K_M or V_{max} of an enzyme or uptake system could influence the rate of metabolite flow through a metabolic pathway, particularly where the enzyme or uptake system is the rate-limiting step in the pathway. For instance, the efficiency of utilization of particular components of fertilizer might be increased by modifying the kinetic properties of the enzyme or uptake system that is rate limiting.

An effective method of obtaining bacterial mutant strains altered in catalytic specificity, K_M, or V_{max} is the selection of cells in a chemostat. The application of chemostat culture to plant cells has been discussed above (Section V,C,1). It is not necessary to resort to chemostat culture, however, to select for mutations in structural genes. Growth in the presence of toxic herbicides or nonmetabolizable substrates can be an effective selective screen provided that a single mutation permits a cell to have a growth advantage. Thus selection for an enzyme with modified specificity or kinetic properties may be achieved simply and directly. Most of the mutations would be expected to be dominant or semidominant. Exceptions would be structural gene mutations that confer resistance to a compound through a lesion in an uptake system or a lesion in an enzyme that converts the compound to a toxic substance.

Studies with bacteria indicate that the evolution of new catalytic specificity in an enzyme may require multiple structural gene mutations in addition to a regulatory mutation. Such insights derived from bacteria should guide the plant somatic cell geneticist away from certain selection schemes. For example, conversion of chlorate to chlorite or other toxic intermediates is mediated by nitrate reductase (de Graff *et al.,* 1973; Oostindier-Braaksma and Feenstra, 1973). Mutations that result in an absence of nitrate reductase activity towards nitrate and chlorate are likely to occur at a reasonable frequency. The inclusion of both nitrate and chlorate in the medium in an effort to select for mutant cells that can discriminate between nitrate and chlorate is a questionable procedure. A mutation that allows nitrate reductase to be active with nitrate, but not chlorate, would be expected to occur at an extremely low frequency, if at all.

VI. OPEN QUESTIONS

Assuming that regeneration of wild-type plant cells and selection of mutant cells are possible, there are additional aspects of cultured plant cells to consider. In this section we examine the effect of mutations on development and the relationship of cultured plant cells to the whole plant.

A. Effect of Mutations Selected *in Vitro* on Plant Development

The question of whether the expression of the mutant phenotype will be observed in regenerated plant tissues is likely to have different answers in diverse experimental systems. Mutant cell lines selected *in vitro* have been regenerated into whole plants and have been shown to possess mutant phenotypes (Carlson, 1970, 1973; Widholm, 1974b; Müller *et al.*, 1977). It seems conceivable that other mutant cell lines will not regenerate into plants carrying the mutant trait. There may be a number of reasons for a failure to regenerate mutant plants. One would be that the mutation is developmentally sensitive, the wild-type function being required for normal development; in this case, no plants would be regenerated from the mutant callus. On the other hand, the regeneration of plants with a wild-type phenotype from putative mutant clones need not point inexorably to the occurrence of an epigenetic event. Reversion of a developmentally sensitive mutation or a secondary suppressor mutation could account for the recovery of wild-type plants. The isolation of conditional mutants or variants, such as temperature-sensitive mutants, would be invaluable in establishing that a specific mutation is incompatible with the course of normal morphogenesis.

B. Relationship of Cultured Plant Cells to Differentiated States of the Whole Plant

Only a limited spectrum of plant proteins and enzymes are expressed in cultured cells. Defined selection schemes for plant cells are limited to those functions that are expressed *in vitro*. The enzymes and proteins that are expressed are subject to unknown selective pressures in cells grown in liquid or agar-solidified medium. In addition, in plants of polyploid ancestry, an enzyme studied in cultured cells may be encoded by a different gene from the same enzyme in a particular differentiated tissue. Thus, the question arises of whether the regulation of protein synthesis observed *in vitro* reflects the regulation that takes place when the cells are part of a differentiated tissue.

VII. CONCLUDING REMARKS

Studies with whole plants and with cultured plant cells may be complementary in elucidating metabolic controls in higher plants. A number of the selection schemes and the methods of analyzing the regulation of protein synthesis used with bacteria can be readily applied to plant cells *in vitro*. Other selection schemes hold promise, but their application to plant cells may prove to be more difficult. Advances made with simpler microbial systems provide insight for researchers working with cultured higher plant cells. It must be recognized, however, that the genetic control mechanisms governing the synthesis of proteins in prokaryotes and eukaryotes are likely to be different. Furthermore, somatic cell genetics of higher plants has a host of unique problems and prospects. Realization of the exciting potential of somatic cell genetics of higher plants requires fresh, imaginative approaches to the problems that limit the recovery of a range of recessive and dominant mutations.

REFERENCES

Afridi, M. M. R. K., and Hewitt, E. J. (1965). *J. Exp. Bot.* **16,** 628–645.
Allard, R. W. (1960). "Principles of Plant Breeding." Wiley, New York.
Bayliss, M. W. (1975). *Chromosoma* **51,** 401–411.
Behrend, J., and Mateles, R. I. (1975). *Plant Physiol.* **56,** 584–589.
Behrend, J., and Mateles, R. I. (1976). *Plant Physiol.* **58,** 510–512.
Bergmann, L. (1960). *J. Gen. Physiol.* **43,** 841–851.
Binns, A., and Meins, F., Jr. (1973). *Proc. Natl. Acad. Sci. U. S. A.* **70,** 2660–2662.
Blakely, L. M., and Steward, F. C. (1964). *Am. J. Bot.* **51,** 780–791.
Bollard, E. G. (1966). *Plant Soil* **25,** 153–166.
Bollard, E. G., Cook, A. R., and Turner, N. A. (1968). *Planta* **83,** 1–12.
Bonner, J. (1976). *In* "Plant Biochemistry" (J. Bonner and J. E. Varner, eds.), 3rd ed., pp. 37–64. Academic Press, New York.
Bradley, M. V. (1954). *Am. J. Bot.* **41,** 398–402.
Braun, A. C. (1959). *Proc. Natl. Acad. Sci. U. S. A.* **45,** 932–938.
Britten, R. J., and Kohne, D. E. (1968). *Science* **161,** 529–540.
Brown, J. E., Brown, P. R., and Clarke, P. H. (1969). *J. Gen. Microbiol.* **57,** 273–295.
Brown, P. R., and Clarke, P. H. (1972). *J. Gen. Microbiol.* **70,** 287–298.
Burk, L. G., Gwynn, G. R., and Chaplin, J. F. (1972). *J. Hered.* **63,** 355–360.
Bussey, H. (1970). *J. Bacteriol.* **101,** 1081–1082.
Cánovas, J. L., Johnson, B. F., and Wheeler, M. L. (1968). *Eur. J. Biochem.* **3,** 305–311.
Carlson, P. S. (1970). *Science* **168,** 487–489.
Carlson, P. S. (1973). *Science* **180,** 1366–1368.
Carlson, P. S., and Polacco, J. C. (1975). *Science* **188,** 622–625.
Cazzulo, J. J., Sundaram, T. K., and Kornberg, H. L. (1969). *Nature (London)* **223,** 1137–1138.
Chaleff, R. S., and Carlson, P. S. (1974a). *Annu. Rev. Genet.* **8,** 267–278.

Chaleff, R. S., and Carlson, P. S. (1974b). *In* "Genetic Manipulations with Plant Material" (L. Ledoux, ed.), pp. 351–363. Plenum, New York.

Chase, S. S. (1969). *Bot. Rev.* **35**, 117–167.

Chu, Y., and Lark, K. G. (1976). *Planta* **132**, 259–268.

Clarke, P. H. (1974). *Symp. Soc. Gen. Microbiol.* **24**, 183–217.

Clarke, P. H., and Lilly, M. D. (1969). *Symp. Soc. Gen. Microbiol.* **19**, 113–159.

Cocking, E. C. (1972). *Annu. Rev. Plant Physiol.* **23**, 29–50.

Cohen-Bazire, G., and Jolit, M. (1953). *Ann. Inst. Pasteur, Paris* **84**, 937–945.

Crawford, I. P. (1975). *Bacteriol. Rev.* **39**, 87–120.

Creasy, L. L., and Zucker, M. (1974). *Recent Adv. Phytochem.* **8**, 1–19.

Das, N. K., Patau, K., and Skoog, F. (1956). *Physiol. Plant.* **9**, 640–651.

Davis, B. D. (1948). *J. Am. Chem. Soc.* **70**, 4267.

Debergh, P., and Nitsch, C. (1973). *C. R. Hebd. Seances Acad. Sci., Ser. D* **276**, 1281–1284.

de Graff, J., Barendsen, W., and Stouthamer, A. H. (1973). *Mol. Gen. Genet.* **121**, 259–269.

Demain, A. L. (1971). *In* "Methods in Enzymology" (W. B. Jakoby, ed.), Vol. 22, pp. 86–95. Academic Press, New York.

Dorée, M., Leguay, J.-J., and Terrine, C. (1972). *Physiol. Veg.* **10**, 115–131.

Dougall, D. K. (1972). *In* "Growth, Nutrition, and Metabolism of Cells in Culture" (G. H. Rothblat and V. J. Cristofalo, eds.), Vol. 2, pp. 372–406. Academic Press, New York.

Drake, J. W. (1970). "The Molecular Basis of Mutation." Holden-Day, San Francisco, California.

Duhamet, L. (1953). *C. R. Seances Soc. Biol. Ses Fil.* **147**, 81–83.

Earle, D. E., and Torrey, J. G. (1965). *Plant Physiol.* **40**, 520–528.

Englesberg, E., and Wilcox, G. (1974). *Annu. Rev. Genet.* **8**, 219–242.

Filner, P. (1965). *Exp. Cell Res.* **39**, 33–39.

Filner, P. (1966). *Biochim. Biophys. Acta* **118**, 299–310.

Filner, P., Wray, J. L., and Varner, J. E. (1969). *Science* **165**, 358–367.

Francis, J. C., and Hansche, P. E. (1972). *Genetics* **70**, 59–73.

Francis, J. C., and Hansche, P. E. (1973). *Genetics* **74**, 259–265.

Frank, E. M., Riker, A. J., and Dye, S. L. (1951). *Plant Physiol.* **26**, 258–267.

Gamborg, O. L., and Shyluk, J. P. (1970). *Plant Physiol.* **45**, 598–600.

Gamborg, O. L., Murashige, T., Thorpe, T. A., and Vasil, I. K. (1976). *In Vitro* **12**, 473–478.

Gautheret, R. J. (1939). *C. R. Hebd. Seances Acad. Sci.* **208**, 118–120.

Gautheret, R. J. (1942). *C. R. Hebd. Seances Acad. Sci.* **214**, 805–807.

Gautheret, R. J. (1955). *Annu. Rev. Plant Physiol.* **6**, 433–484.

Gehring, W. (1968). *In* "The Stability of the Differentiated State" (H. Ursprung, ed.), pp. 136–154. Springer-Verlag, Berlin and New York.

Gibbs, J. L., and Dougall, D. K. (1965). *Exp. Cell Res.* **40**, 85–95.

Goodspeed, T. H. (1954). *Chron. Bot.* **16**, 1–536.

Greenberg, D. M. (1969). *Metab. Pathways, 3rd Ed.* **3**, 237–315.

Gresshoff, P. M., and Doy, C. H. (1972a). *Aust. J. Biol. Sci.* **25**, 259–264.

Gresshoff, P. M., and Doy, C. H. (1972b). *Planta* **107**, 161–170.

Grisebach, H., and Hahlbrock, K. (1974). *Recent Adv. Phytochem.* **8**, 21–52.

Gross, S. R. (1969). *Annu. Rev. Genet.* **3**, 395–424.

Guha, S., and Maheshwari, S. C. (1964). *Nature (London)* **204**, 497.

Guha, S., and Maheshwari, S. C. (1966). *Nature (London)* **212**, 97–98.

Guha, S., and Maheshwari, S. C. (1967). *Phytomorphology* **17**, 454–461.

Hahlbrock, K., and Schröder, J. (1975). *Arch. Biochem. Biophys.* **166**, 47–53.

Halperin, W. (1966). *Am. J. Bot.* **53**, 443–453.

Hamkalo, B. A., and Miller, O. L., Jr. (1973). *Annu. Rev. Biochem.* **42**, 379–396.
Hansche, P. E. (1975). *Genetics* **79**, 661–674.
Hartley, B. S. (1974). *Symp. Soc. Gen. Microbiol.* **24**, 151–182.
Hegeman, G. D. (1966a). *J. Bacteriol.* **91**, 1140–1154.
Hegeman, G. D. (1966b). *J. Bacteriol.* **91**, 1161–1167.
Heimer, Y. M., and Filner, P. (1970). *Biochim. Biophys. Acta* **215**, 152–165.
Heimer, Y. M., and Filner, P. (1971). *Biochim. Biophys. Acta* **230**, 362–372.
Heimer, Y. M., Wray, J. L., and Filner, P. (1969). *Plant Physiol.* **44**, 1197–1199.
Helgeson, J. P. S., Krueger, S. M., and Upper, C. D. (1969). *Plant Physiol.* **44**, 193–198.
Henshaw, G. G., Jha, K. K., Mehta, A. R., Shakeshaft, D. J., and Street, H. E. (1966). *J. Exp. Bot.* **17**, 362–377.
Hewitt, E. J., Hucklesby, D. P., and Notton, B. A. (1976). *In* "Plant Biochemistry" (J. Bonner and J. E. Varner, eds.), pp. 633–681. Academic Press, New York.
Hill, C. W., Foulds, J., Soll, L., and Berg, P. (1969). *J. Mol. Biol.* **39**, 563–581.
Horiuchi, T., Horiuchi, S., and Novick, A. (1962). *Genetics* **48**, 157–169.
Ingle, J., Joy, K. W., and Hageman, R. H. (1966). *Biochem. J.* **100**, 577–588.
Jackson, E. N., and Yanofsky, C. (1973). *J. Bacteriol.* **116**, 33–40.
Jacob, F., and Monod, J. (1961). *J. Mol. Biol.* **3**, 318–358.
Jobe, A., and Bourgeois, S. (1972). *J. Mol. Biol.* **69**, 397–408.
Jones, E. L., Hildebrandt, A. C., Riker, A. J., and Wu, J. H. (1960). *Am. J. Bot.* **47**, 468–475.
Jouanneau, J. P., and Péaud-Lenoël, C. (1967). *Physiol. Plant.* **20**, 834–850.
Kao, K. N., Miller, R. A., Gamborg, O. L., and Harvey, B. L. (1970). *Can. J. Genet. Cytol.* **12**, 297–301.
Kasperbauer, M. J., and Collins, G. B. (1972). *Crop Sci.* **12**, 98–101.
Kelker, H. C., and Filner, P. (1971). *Biochim. Biophys. Acta* **252**, 69–82.
Kemp, M. B., and Hegeman, G. D. (1968). *J. Bacteriol.* **96**, 1488–1499.
Kepes, A., and Cohen, G. N. (1962). *In* "The Bacteria" (I. C. Gunsalus and R. Y. Stanier, eds.), Vol. 4, pp. 179–221. Academic Press, New York.
Key, J. L. (1976). *In* "Plant Biochemistry" (J. Bonner and J. E. Varner, eds.), pp. 463–505. Academic Press, New York.
Lamport, D. T. A. (1964). *Exp. Cell Res.* **33**, 195–206.
Lederberg, J., and Lederberg, E. M. (1952). *J. Bacteriol.* **63**, 399–406.
Lederberg, J., and Zinder, N. (1948). *J. Am. Chem. Soc.* **70**, 4267.
Li, S. L., Rédei, G. P., and Gowans, C. S. (1967). *Mol. Gen. Genet.* **100**, 77–83.
Linsmaier, E. M., and Skoog, F. (1965). *Physiol. Plant.* **18**, 100–127.
Luria, S. E. (1960). *In* "The Bacteria" (I. C. Gunsalus and R. Y. Stanier, eds.), Vol. 1, pp. 1–34. Academic Press, New York.
Lutz, A. (1971). *Colloq. Int. C. N. R. S.* **193**, 163–168.
Mackenzie, I. A., Konar, A., and Street, H. E. (1972). *New Phytol.* **71**, 633–638.
Mahadevan, P. R., and Eberhart, B. M. (1962). *J. Cell. Comp. Physiol.* **60**, 281–284.
Márton, L., and Maliga, P. (1975). *Plant Sci. Lett.* **5**, 77–81.
Matsumoto, H., Yasuda, T., Kobayashi, M., and Takahashi, E. (1966). *Soil Sci. Plant Nutr. (Tokyo)* **12**, 33–38.
Melchers, G. (1971). *Colloq. Int. C. N. R. S.* **193**, 229–234.
Melchers, G., and Bergmann, L. (1958). *Ber. Dtsch. Bot. Ges.* **71**, 459–473.
Miflin, B. J. (1973). *In* "Biosynthesis and Its Control in Plants" (B. V. Milborrow, ed.), pp. 49–68. Academic Press, New York.
Monod, J. (1942). "Recherches sur la croissance des cultures bactériennes." Hermann, Paris.

Muir, W. H., Hildebrandt, A. C., and Riker, A. J. (1954). *Science* **119**, 877–878.
Muir, W. H., Hildebrandt, A. C., and Riker, A. J. (1958). *Am. J. Bot.* **45**, 589–597.
Müller, A. J., Grafe, R., Mendel, R. R., and Saalbach, I. (1977). *In* "Experimental Mutagenesis in Plants," Proc. Symp. held in Varna, October 1976, Sofia 1977 (in press).
Murashige, T. (1973). *In Vitro* **9**, 81–85.
Murashige, T. (1974). *Annu. Rev. Plant Physiol.* **25**, 135–166.
Murashige, T., and Nakano, R. (1967). *Am. J. Bot.* **54**, 963–970.
Murashige, T., and Skoog, F. (1962). *Physiol. Plant.* **15**, 473–497.
Nanney, D. L. (1958). *Proc. Natl. Acad. Sci. U. S. A.* **44**, 712–717.
Narayanaswamy, S. (1977). *In* "Plant Cell, Tissue, and Organ Culture" (J. Reinert and Y. P. S. Bajaj, eds.), pp. 179–248. Springer-Verlag, Berlin and New York.
Nickell, L. G. (1956). *Proc. Natl. Acad. Sci. U. S. A.* **42**, 848–850.
Nickell, L. G., and Heinz, D. J. (1973). *In* "Genes, Enzymes and Populations" (A. M. Srb, ed.), pp. 109–128. Plenum, New York.
Nishi, A., and Sugano, N. (1970). *Plant Cell Physiol.* **11**, 757–765.
Nitsch, C. (1974). *In* "Haploids in Higher Plants: Advances and Potentials" (K. J. Kasha, ed.), pp. 123–135. University of Guelph, Guelph, Canada.
Nobécourt, P. (1939). *C. R. Seances Soc. Biol. Ses Fil.* **130**, 1270–1271.
Northcote, D. H. (1969). *Symp. Soc. Gen. Microbiol.* **19**, 333–349.
Northrop, J. H., Kunitz, M., and Herriot, R. (1948). "Crystalline Enzymes." Columbia Univ. Press, New York.
Novick, A., and Horiuchi, T. (1961). *Cold Spring Harbor Symp. Quant. Biol.* **26**, 239–245.
Novick, A., and Weiner, M. (1957). *Proc. Natl. Acad. Sci. U. S. A.* **43**, 553–566.
Ohyama, K. (1974). *Exp. Cell Res.* **89**, 31–38.
Okamura, S., Sueki, K., and Nishi, A. (1975). *Physiol. Plant.* **33**, 251–255.
Oostindier-Braaksma, F. J., and Feenstra, W. J. (1973). *Mutat. Res.* **19**, 175–185.
Ornston, L. N. (1971). *Bacteriol. Rev.* **35**, 87–116.
Ornston, L. N., and Parke, D. (1977). *Curr. Top. Cell. Regu.* **12**, 209–262.
Ornston, L. N., Ornston, M. K., and Chou, G. (1969). *Biochem. Biophys. Res. Commun.* **36**, 179–184.
Palleroni, N. J., and Stanier, R. Y. (1964). *J. Gen. Microbiol.* **35**, 319–334.
Palmer, J. E., and Widholm, J. (1975). *Plant Physiol.* **56**, 233–238.
Parke, D., and Ornston, L. N. (1976). *J. Bacteriol.* **126**, 272–281.
Patau, K., Das, N. K., and Skoog, F. (1957). *Physiol. Plant.* **10**, 949–966.
Polacco, J. C. (1976). *Plant Physiol.* **58**, 350–357.
Powell, E. O. (1958). *J. Gen. Microbiol.* **18**, 259–268.
Pressey, R., and Shaw, R. (1966). *Plant Physiol.* **41**, 1657–1661.
Puck, T. T., and Kao, F. T. (1967). *Proc. Natl. Acad. Sci. U. S. A.* **58**, 1227–1234.
Puck, T. T., and Marcus, P. I. (1955). *Proc. Natl. Acad. Sci. U. S. A.* **41**, 432–437.
Rasse-Messenguy, F., and Fink, G, R. (1973). *In* "Genes, Enzymes and Populations" (A. M. Srb, ed.), pp. 85–95. Plenum, New York.
Raveh, D., Huberman, E., and Galun, E. (1973). *In Vitro* **9**, 216–222.
Reinert, J. (1973). *In* "Plant Tissue and Cell Culture" (H. E. Street, ed.), pp. 338–355. Blackwell, Oxford.
Reinert, J., and Kuster, H.-J. (1966). *Z. Pflanzenphysiol.* **54**, 213–222.
Rigano, C. (1971). *Arch. Microbiol.* **76**, 265–276.
Rigby, P. W. J., Burleigh, B. D., and Hartley, B. S. (1974). *Nature (London)* **251**, 200–204.
Riker, A. J., and Gutsche, A. E. (1948). *Am. J. Bot.* **35**, 227–238.
Rodwell, V. W. (1969). *Metab. Pathways 3rd Ed.* **3**, 317–373.
Sacristán, M. D., and Melchers, G. (1969). *Mol. Gen. Genet.* **105**, 317–333.

Salisbury, F. B., and Ross, C. (1969). "Plant Physiology." Wadsworth, Belmont, California.
Schimke, R. T., and Doyle, D. (1970). *Annu. Rev. Biochem.* **39**, 929–976.
Schimke, R. T., Sweeney, E. W., and Berlin, C. M. (1965). *J. Biol. Chem.* **240**, 322–331.
Schröder, J., Betz, B., and Hahlbrock, K. (1977). *Plant Physiol.* **60**, 440–445.
Schulte, U., and Zenk, M. H. (1977). *Physiol. Plant.* **39**, 139–142.
Schwartz, D. (1966). *Proc. Natl. Acad. Sci. U. S. A.* **56**, 1431–1436.
Sharp, W. R., Dougall, D. K., and Paddock, E. F. (1971). *Bull. Torrey Bot. Club* **98**, 219–222.
Sharp, W. R., Raskin, R. S., and Sommer, H. E. (1972). *Planta* **104**, 357–361.
Sheridan, W. F. (1974). *In* "Genetic Manipulations with Plant Material" (L. Ledoux, ed.), pp. 263–295. Plenum, New York.
Shimada, T., and Tabata, M. (1967). *Jpn. J. Genet.* **42**, 195–201.
Smith, H. H. (1974). *Bioscience* **24**, 269–276.
Stanier, R. Y., Palleroni, N. J., and Doudoroff, M. (1966). *J. Gen. Microbiol.* **43**, 159–271.
Stanier, R. Y., Adelberg, E. A., and Ingraham, J. (1976). "The Microbial World." Prentice-Hall, Englewood Cliffs, New Jersey.
Stephenson, M. (1949). "Bacterial Metabolism." Longman, Green, New York.
Stewart, G. R. (1972). *J. Exp. Bot.* **23**, 171–183.
Street, H. E. (1973a). *In* "Biosynthesis and its Control in Plants" (B. V. Milborrow, ed.), pp. 93–125. Academic Press, New York.
Street, H. E. (1973b). *In* "Plant Tissue and Cell Culture" (H. E. Street, ed.), pp. 191–204. Blackwell, Oxford.
Stuart, R., and Street, H. E. (1971). *J. Exp. Bot.* **22**, 96–106.
Sunderland, N. (1971). *Sci. Prog. (Oxford)* **59**, 527–549.
Sunderland, N. (1973a). *In* "Plant Tissue and Cell Culture" (H. E. Street, ed.), pp. 161–190. Blackwell, Oxford.
Sunderland, N. (1973b). *In* "Plant Tissue and Cell Culture" (H. E. Street, ed.), pp. 205–239. Blackwell, Oxford.
Sung, Z. R. (1976). *Genetics* **84**, 51–57.
Torrey, J. G., and Reinert, J. (1961). *Plant Physiol.* **36**, 483–490.
Umbarger, H. E. (1969). *Annu. Rev. Biochem.* **38**, 323–370.
Umbarger, H. E. (1971). *Adv. Genet.* **16**, 119–140.
Varner, J. E., and Ho, D. T.-H. (1976). *In* "Plant Biochemistry" (J. Bonner and J. E. Varner, eds.), pp. 713–770. Academic Press, New York.
Vasil, I. K. (1973). *Naturwissenschaften* **60**, 247–253.
Vasil, I. K., and Vasil, V. (1972). *In Vitro* **8**, 117–127.
Vega, J. Ma., Herrera, J., Aparicio, P. J., Paneque, A., and Losada, M. (1971). *Plant Physiol.* **48**, 294–299.
Vysotskaya, E. F., and Gamburg, K. Z. (1975). *Sov. Plant Physiol. (Engl. Transl.)* **22**, 47–52.
Welander, T. (1976). *Physiol. Plant* **36**, 7–10.
White, P. R. (1939). *Am. J. Bot.* **26**, 59–64.
White, P. R. (1943). *Growth* **7**, 53–65.
Widholm, J. M. (1972a). *Biochim. Biophys. Acta* **261**, 44–51.
Widholm, J. M. (1972b). *Biochim. Biophys. Acta* **261**, 52–58.
Widholm, J. M. (1972c). *Biochim. Biophys. Acta* **279**, 48–57.
Widholm, J. M. (1972d). *Stain Technol.* **47**, 189–194.
Widholm, J. M. (1974a). *Plant Sci. Lett.* **3**, 323–330.

Widholm, J. M. (1974b). *In* "Tissue Culture and Plant Science" (H. E. Street, ed.), pp. 287–299. Academic Press, New York.

Widholm, J. M. (1976). *Can. J. Bot.* **54,** 1523–1529.

Widholm, J. M. (1977). *Crop Sci.* **17,** 597–600.

Wray, J. L., and Filner, P. (1970). *Biochem. J.* **119,** 715–725.

Zenk, M. H. (1974). *In* "Haploids in Higher Plants: Advances and Potentials" (K. J. Kasha, ed.), pp. 339–353. University of Guelph, Guelph, Canada

Zielke, H. R., and Filner, P. (1971). *J. Biol. Chem.* **246,** 1772–1779.

7

Genetic Polymorphism among Enzyme Loci

GEORGE JOHNSON

I. INTRODUCTION

While intermediary metabolism has been the subject of intensive investigation for many years, relatively little is known about genetic variation affecting such processes in natural populations. What we do know tends often to be restricted to studies of individual enzymes, with little informa-

PHYSIOLOGICAL GENETICS

tion on how variation at these specific loci interacts with that of other loci whose enzymes catalyze reactions in the same or related sequences. From a broader genetic perspective, however, this is just the sort of information one would like to have. There are a variety of empirical approaches now being employed in attempts to address such multilocus problems. One approach, coordinate physiological and genetic analysis of discrete metabolic phenotypes, seems to me particularly promising. This chapter reviews the available information on enzyme polymorphism among loci mediating important metabolic processes and discusses the methodological and empirical requirements of a physiologically oriented multilocus study.

II. LEVELS OF ENZYME POLYMORPHISM

A. Initial Estimates

In the first reported studies of naturally occurring enzyme polymorphism detected by electrophoretic comparisons, the level of heterozygosity was estimated at 0.12 in *Drosophila* (Lewontin and Hubby, 1966) and 0.10 in man (Harris, 1966). This result has proven remarkably general among the large number of subsequent surveys of levels of genetic variation in nature. Levels of variation are usually reported as being in the range of 10 to 20% average heterozygosity, with about one-third of the loci examined in large studies proving to have at least two alleles in frequencies greater than 1%. Recent exhaustive reviews are available of the survey analyses of enzyme polymorphism among animal (Powell, 1975) and plant (Hamrick, 1978) populations.

As an example of the constancy of the patterns of variation seen at different enzyme loci in these initial studies, Table I presents data on average heterozygosity in man, small vertebrates, *Drosophila,* and the higher plant *Silene maritima.* Certain loci seem consistently more variable than others, but the pattern across very different organisms is quite remarkably similar: The same sorts of loci vary, and heterozygosity varies over the same range of 10 to 20%. Thus amounts of variation compared across widely divergent taxonomic groups differ little more than what one might expect comparing divergent species within a single genus. Compare these results to those of Table II, which presents data on the ecologically diverse butterfly genus *Colias:* The range in levels of variability is as great within the one genus as within the diverse collection of organisms of Table I. This is not to say that there are not clear exceptions, cases of very little or extraordinarily abundant variation. The general re-

TABLE I Levels of Enzyme Polymorphism in Diverse Species[a]

Enzyme	Man	Small vertebrates	Drosophila williston	Silene maritima
Peptidases	0.08	0.17	0.64	0.46
Acid phosphatase	0.17	—	0.14	0.65
Esterases	0.05	0.30	0.26	High
Alcohol dehydrogenase (AdH)	0.48, 0.07, 0	0.21	0.11	0.49
Adenylate kinase (AK)	0.09	—	0.25	0.39
Glucose-6-phosphate dehydrogenase (G6PdH)	—	0.03	—	—
Hexokinase (HK)	0.05, 0	—	0.05	—
Malic enzyme (ME)	0.30, 0	0.20	0.09	—
Phosphoglucoisomerase (PGI)	0	0.16	—	—
Phosphoglucomutase (PGM)	0.36, 0, 0.38	0.14, 0.24	0.18	—
Aldolase (ALD)	0	—	0.14	0
α-Glycerophosphate dehydrogenase (α-GPdH)	0	0.10	0.03	—
Isocitrate dehydrogenase (IdH)	0	0.01, 0.13	0.06	0
Malate dehydrogenase (MdH)	0, 0	0.04, 0.07	0.06	0.50 ?
6-Phosphogluconate dehydrogenase	0	0.11	—	0.13
Triose phosphate isomerase (TPI)	0	—	0.04	—
Glutamate-oxalacetate transaminase (GOT)	0, 0.03	0.08, 0.01	—	—
Lactate dehydrogenase (LdH)	0	0.03, 0.09	—	—
Mean heterozygosity (HET)	0.07	0.12	0.18	0.13

[a] Values are from Johnson, 1974; Powell, 1975; Hammrick, 1978.

TABLE II Levels of Enzyme Polymorphism in Species on the Genus Colias[a]

Locus[b]	C. philodice		C. scudderi	C. alexandra		C. meadii		
	Slate River (montane)	Hotchkiss (lowland)	Cement Creek (montane)	East River (montane)	Cement Creek (montane)	Cumberland (alpine)	Mesa Seco (alpine)	Mesa Seco (montane)
α-GPdH	0.40	0.10	0	0.17	0.15	0.05	0.11	0.43
G6PdH	0	0.05	0.25	0.35	0.38	0.08	0.55	0.85
MdH-1	0.10	0	0	0.05	0	0	0	0
MdH-2	0.85	0.67	0	0.30	0.38	0.13	0.15	0.45
ME	0	0.20	0.25	0	0.08	0	0.05	0.15
FUM	0.11	0.45	0.20	0.05	0.08	0	0	0.55
PGM	0.45	0.35	0.10	0.11	0.18	0.28	0.65	0.65
TPI	0.35	0.15	0.25	0.20	0	0.15	0.05	0.05
EST-1	0	0	0	0	0	0.05	0	0
EST-2	0.65	0.50	0.15	—	—	0.20	—	0
HET	0.29	0.25	0.12	0.14	0.14	0.09	0.17	0.31

[a] Values from Johnson, 1976c. Data are presented as observed heterozygosity.

[b] α-GPdH, α-glycerophosphate dehydrogenase locus: G6PdH, glucose-6-phosphate dehydrogenase locus: MdH-1 and MdH-2, malate dehydrogenase loci; ME, malic enzyme locus: FUM, fumarase locus; PGM, phosphoglucomutase locus; TPI triose phosphate isomerase locus; EST-1 and EST-2, esterase loci; HET, mean heterozygosity.

sult, however, seems clear: Levels of observed heterozygosity are quite high, and of the same order of magnitude, in most diploid organisms whose natural populations have been studied.

B. Hidden Heterogeneity

Within the past few years a variety of studies of electrophoretically detected enzyme variation in insects have suggested the presence of additional variant classes not normally resolved by standard electrophoretic techniques. An early suggestion of such cryptic variation was based on the observation that when variants were analyzed on carefully standardized gels, there was more variation than predicted by experimental error (Johnson, 1971a). The first clear examples of cryptic variation were not, however, electrophoretic, but rather involved differential heat stability. In the last three years a variety of techniques (heat inactivation, changing gel porosity, changing gel pH, etc.) have revealed a large previously undetected amount of genetic polymorphism at enzyme loci (Bernstein *et al.*, 1973; Singh *et al.*, 1974, 1975, 1976; Thorig *et al.*, 1975; Johnson, 1975a, 1976b, 1977b; Trippa *et al.*, 1976; Milkman, 1976; Cochrane, 1976; Coyne, 1976). The proper genetic interpretation of these data remains uncertain and is discussed in detail in Section III.

C. The "Selection/Neutrality" Argument

The occurrence of high levels of variation at enzyme loci came as somewhat of a surprise, and it has engendered a lively discussion among population geneticists as to why so much is seen. While many interpretations are possible, much of the discussion has centered upon two contrasting views: (1) That the bulk of the electrophoretically detected variation is there because it is actively maintained by selection and (2) that most electrophoretically detected variants are not functionally different (and thus are equivalent, or "neutral," in their sensitivity to selection), and are common due to the action of other, perhaps random, factors. The matter has been extensively reviewed (Harris, 1971; Johnson, 1973a; Lewontin, 1974; Ewens, 1977).

Much of the experimental work has proved ambiguous, with alternative interpretations equally consistent with the data. The issue thus remains an open one, with individual investigators often adopting polarized viewpoints. If nothing else, these studies have served to illustrate the very real difficulty in formulating a series of rejectable alternative hypotheses concerning this issue, such that experiments might be designed in a straightforward manner.

The avenue which seems to me most promising in this regard is suggested by the observation that although the variation we are concerned with is at enzyme loci, relatively few attempts have been made to examine directly the phenotypes which these enzymes modulate. If enzyme polymorphism is maintained by direct selection, then it is these phenotypes upon which selection must act. If the functional differences between alternative alleles at a locus act to produce the high levels of variation that we see, then it can only be because of their physiological effect upon the individuals, which after all are the targets of selection. Conversely, if the high levels of variation do not reflect functional difference, but rather come about because of linkage to other genes under selection, or because of some other "random" factor not determined by the enzyme variation, then the variation must be considered neutral to selection in any meaningful evolutionary sense. (No nucleotide change is ever without some effect in the sense that it alters future possibilities of change.) The key distinction is that of function. When phrased in this manner, the question becomes largely a physiological one and leads to hypotheses concerning selection which are formulated in terms of specific phenotypic differences. Clearly rejectable hypotheses ought to be more easily arrived at within such an experimental approach.

III. DETECTING AND CHARACTERIZING VARIATION

A. Gel Sieving Analysis

Alleles of electrophoretically detected enzyme loci are typically characterized by comparing their mobilities on starch or acrylamide gels. Over the last decade this has proved a particularly straightforward and convenient approach to detect genetic polymorphism, since mobility classes appear clearly discontinuous in a wide range of mammalian, plant, and insect systems (Harris, 1966; Hamrick, 1978; Powell, 1975). Within the past few years, a variety of studies of electrophoretically detected enzyme variation in insects have suggested the presence of additional variant classes not normally resolved by standard electrophoretic techniques. Thus when variants are analyzed on carefully standardized gels, there is more variation in mobility than predicted by experimental error (Johnson, 1971a; Petrakis and Brown, 1970): When two internal standard proteins are routinely run along with the sample in each gel migration path, the gel to gel variation in enzyme mobility is much greater than that of the standards. An early comparison of the heat stability of variants at the *esterase-6* locus of *Drosophila melanogaster* suggested the existence of alleles not distinguishable by electrophoresis (Wright and MacIntyre, 1965). More recently, when individuals of a presumptively homogeneous line of octanol dehydrogenase in *D. pseudoobscura* are analyzed electrophoretically and the gels incubated at high temperature, some individuals show more stability than others (Bernstein *et al.*, 1973); subsequent studies have suggested that such variation in heat stability is heritable (Singh *et al.*, 1974, 1975). In the few years since these initial reports, a variety of studies have confirmed the existence of heat-stability variants in *Drosophila:* phosphoglucomutase (PGM) (Trippa *et al.*, 1976), alcohol dehydrogenase (AdH) (Milkman, 1976), and esterase (Cochrane, 1976). A particularly careful analysis of thermal variants of alcohol dehydrogenase in *D. melanogaster* (Thorig *et al.*, 1975; Sampsell, 1977) confirmed in this instance that the heat-labile variant mapped to the same genetic locus as the *AdH* structural gene.

Continued analysis of interindividual variation on standardized gels subsequently revealed cryptic variation of a very different sort. Investigations of small differences in mobility have been carried out in my laboratory using careful standardization in order to ensure gel to gel reproducibility. Rigorous standardization may be achieved by running two internal standard proteins along with the sample in each gel migration path. This standardization technique reveals the existence of small-scale, but reproducible, differences in gel mobility; gel-to-gel variation between wild-caught individuals is much greater than what experimental error would have produced. It was concluded that this newly detected variation is a property of the proteins themselves, rather than error, and that it represents either posttranslational modifications or "an unresolved distribution of alleles with minor electrophoretic mobility differences" (Johnson, 1971a).

In principle, the fine-scale variation in electrophoretic mobility detected in rigorously standardized gels might reflect either partial differences in charge (perhaps the amino acid residue in question is only partially exposed) or differences in conformation. To differentiate clearly between these alternatives requires a more direct characterization of the alleles. The mobility of individuals on 7% acrylamide gels is not adequate for this purpose because discrete discontinuous classes are not obtained. What is required is a direct characterization of those properties of shape and charge which are responsible for producing the observed variation in electrophoretic mobility.

Such a characterization is by no means difficult to obtain, at least in principle. The theory describing how a protein migrates in polyacrylamide gel electrophoresis stems from the observation by Ferguson that a protein's mobility in gel electrophoresis is a logarithmic function of gel pore size (Ferguson, 1964). This suggests a straightforward theoretical description

$$R_f = (M_0/u_f) \exp(K_r T) \tag{1}$$

where R_f is the protein mobility relative to front, u_f is a buffer constant, M_0 the free electrophoretic mobility, which is a function of net charge, K_r is the retardation coefficient, which is a function of molecular weight and protein conformation, and T is the percent acrylamide which determines pore size and is inversely proportional to it. In this model, the rate of migration is seen to vary not only as a function of net charge (expressed as free electrophoretic mobility M_0 corrected by a constant u_f for the buffer employed), but also, as suggested by Ferguson, as a logarithmic function of the gel pore size (expressed at per cent acrylamide) and of the gel–protein interaction as the protein passes through these pores (expressed as the retardation coefficient K_r).

The approach suggested by Eq. (1) seems ideally suited to the experimental problem posed by the allelic variation in mobility (R_f) discussed above, as it permits independent characterization of the contributions of charge and of interactive effects such as size or conformational differences to the electrophoretic behavior of a protein. These parameters may be empirically estimated by the simple expedient of taking the logarithm of both sides of Eq. (1), yielding a function of linear form

$$\underbrace{\log R_f}_{y} = \underbrace{\log (M_0/u_f)}_{\text{intercept}} + \underbrace{K_r}_{\text{slope}} \times \underbrace{T}_{x} \tag{2}$$

One runs replicate samples of an individual in parallel on several gels of differing pore size T and determines for each gel the corresponding mobility R_f, thus characterizing directly the degree to which reducing the pore size retards migration. Regressing $\log R_f$ on T, one obtains a linear plot with a slope of K_r and an intercept whose antilog is M_0 divided by a constant. A typical result is presented in Fig. 1.

When gel sieving analysis such as described above was carried out on individuals sampled from natural populations of butterflies, it was immediately apparent that the variation in mobility observed previously reflected more than simple charge differences. If the only source of difference is net charge (presumably produced by amino acid substitutions involving charged residues), then one would expect variants to have similar retardation coefficients K_r and to differ primarily in free electrophoretic mobility M_0. The range of variability in their K_r values would be expected to be limited to about that observed for the hemoglobin internal standard. In fact, the distribution of K_r values obtained is very much

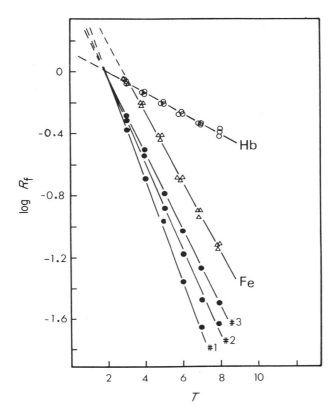

Fig. 1. Gel sieving behavior of three fumarase variants of *Colias meadii* (Johnson, 1977b). Hb (hemoglobin) and Fe (ferritin) are two internal standard proteins run in the same gels.

broader than the corresponding hemoglobin distribution (Johnson, 1976a). Thus differences other than charge clearly contribute to the differences seen in mobility on 7% acrylamide gels.

This proved an important result, as it provides the basis for understanding the intergel variation in mobility consistently seen in samples of natural populations. The difficulty in analysis on 7% acrylamide gels is due to the fact that charge and size/conformation *interact* in determining mobility, and these two protein properties prove to vary concordinately—as M_0 values increase, so do the absolute values of K_r (e.g., bigger or more asymmetric proteins have greater net charge). The result is that the mobility functions described by Eq. (2) intersect at intermediate gel pore sizes. The nature of the mobility variation observed earlier is now clear (Fig. 2). A survey conducted at 5% acrylamide (equivalent to 10–11% starch) does not discriminate between variants, and reveals a single uniform mobility type. Such a survey would classify this locus as uniformly homozygous. A survey conducted at 7% acrylamide, as were my previous surveys, would report two segregating alleles (see Johnson, 1976c), with considerable variation in the exact mobility observed. This is the mobility variation discussed above, and it reflects the fact that there are a minimum of five alleles segregating at this locus.

Gel sieving analysis thus provides direct evidence of protein heterogeneity within electrophoretic classes. Note also that there are alleles which do not differ in net charge, but differ only in K_r.

Experiments conducted in my laboratory in the last several years have focused on the genetic properties of variants detected by gel sieving. Early studies indicated that hidden variants of α-glycerophosphate dehydrogenase (α-GPdH) in *Colias* segregate in crosses (Johnson, 1976a), suggesting the possibility that the variants might be allelic. The pattern of extensive hidden variation proved not to be unique to this locus. In a survey of 14 enzyme loci in *Colias* (Johnson, 1977b), for all loci but malate dehydrogenase (MdH) several common variants are detected which cannot be distinguished from one another on routine 7% acrylamide gels. In addition, other variants occur at low frequency at all the loci examined. Fully 70% of all variants differ significantly in K_r (and thus presumably in shape); 10% differ *only* in K_r. This result is particularly significant in light

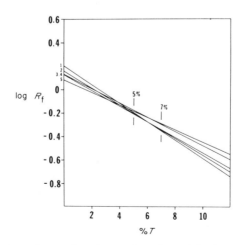

Fig. 2. Five allelic variants of α-glycerophosphate dehydrogenase in *Colias*. Because more asymmetric variants are more highly charged in alkaline buffer, the lines cross at intermediate pore sizes and thus are not discriminated from one another within that range.

of the result presented in Fig. 5. The frequency distribution of the variants in natural populations is very skewed, with one or a few variants occurring commonly at a locus, and the others rarely or uniquely. Very notably, these studies revealed a significant deficiency of observed heterozygotes at all loci examined.

Data have recently been reported for hidden variants at the xanthine dehydrogenase (XdH) locus in *Drosophila pseudoobscura* (Singh *et al.*, 1976) which are in my judgment in broad agreement with the results discussed above: Instead of estimating R_f at six gel concentrations and evaluating the function, two dissimilar gel concentrations are simply compared for R_f difference; instead of estimating isoelectric point, two buffer pHs are compared for their effect on R_f.

The implications of these findings may mean that the degree of polymorphism in all creatures has been grossly underestimated. In order to evaluate the impact of this notion, it is paramount that we learn to what extent this variation is truly allelic. It might in principle represent nucleotide substitutions in the structural gene, or nongenetic posttranslational modifications of the protein gene product (which would not be heritable), or posttranslational modifications mediated by the gene product of another, second site, locus (which would be heritable), or all three mechanisms may be operative.

B. Genetic versus Epigenetic Variation

Current evidence strongly suggests that the bulk of the variation detected by gel sieving analyses is heritable. In my laboratory we have pursued three lines of investigation.

1. Variability in Inbred Lines of *Drosophila*

Of central importance is the demonstration that the large amounts of variation detected by gel sieving analysis do not result from experimental error or some uncontrolled physiological variable. These possibilities may be ruled out by testing for variation in inbred lines of *Drosophila*. In all cases, the variation within inbred lines is no greater than that of the hemoglobin standard run in the same gels. These results thus indicate that gel sieving variation is a phenomenon which does not occur in the absence of genic variation and suggest a heritable basis for the variation.

2. Controlled Crosses

Previous crosses carried out in *Colias* butterflies have indicated that gel sieving variants at the α-GPdH locus segregate in the F_2 generation in Mendelian proportions. Crosses carried out in *Drosophila pseudoobscura* also support a genetic interpretation. When a variant is crossed with itself, only this variant type is found among the F_2 progeny. However, when two variants are crossed with each other, both variant types segregate among the F_2 progeny (Johnson, 1977c).

3. Progeny Tests

The above results are all consistent with a heritable basis to the gel sieving variation. No crosses, however, have been carried out on the rare variant types, whose genetic nature is of particular interest. In order to assess the heritability of such variants, direct progeny tests were carried out of variants at four esterase loci of the ocean fish *Zoarces,* in collaboration with F. Christiansen and V. Simonsen of Aarhus University, Denmark. The fish is viviparous, each gravid female carrying up to 50 live progeny. Thus, a survey of female fish permits one to carry out a direct progeny test of any variants detected by gel sieving analysis. A total of 80 adult fish were surveyed, the four loci being analyzed in each fish. Each of the four loci exhibited several variant classes. Every detected variant proved heritable. These results thus establish that the preponderance of esterase variation detected by gel sieving in *Zoarces* is a heritable variation. As the pattern of variation is not substantially different from that seen for *Drosophila* or *Colias,* these results suggest that gel sieving variants in general will prove heritable.

The large and not satisfactorily explained deficiency of heterozygotes among surveys of gel sieving variation in natural populations (Johnson, 1977b) coupled with the evidence above that much of this variation is heritable, suggests that in the samples surveyed by this author there may be widespread posttranslational modification resulting from the action of one or more second-site polymorphic loci. The action of several such modifying loci might explain the highly skewed frequency distribution, with typically a few common and many rare types. In order to investigate whether such second-site modifiers would be expected to produce the sorts of gel sieving variation seen in my results, the author, in collaboration with Dr. V. Finnerty, has undertaken an examination of variation at the *XdH* locus of *D. melanogaster* (Finnerty and Johnson, 1979). This system is of particular interest because (1) Lewontin and co-workers have shown the *XdH* locus in *Drosophila* to contain considerable hidden variation, much of which appears heritable (Singh *et al.,* 1976); (2) the system is very well characterized in terms of both genetic and biochemical studies of the structural locus for XdH (52.3 on the chromosome 3) (Chovnick *et al.,* 1977). In addition, three other distinct loci are known to affect both XdH and aldehyde oxidase (AO) activity (Finnerty, 1976). One such locus, maroonlike (*ma-l*), unlinked to either of these structural loci (66 on the X chromosome), is nevertheless required for each activity. My studies with Finnerty thus far indicate that *ma-l* mediates a posttranslational modification of these proteins: Examination by gel sieving of a series of *ma-l* heteroallelic combinations indicates that in every case there is a significant alteration in protein shape (Fig. 3). The magnitude of the posttranslational K_r differences in XdH and AO are consistent with the differences reported in earlier gel sieving surveys. The implication is that heritable hidden variation at this locus, and potentially at other loci, may not be allelic to the structural gene. Such a hypothesis would explain the heterozygote deficiency among heritable "alleles" sampled from natural populations: posttranslational modifiers would act as simple dominants and would not be expected to produce intermediate "heterozygote" bands (see case 3, Fig. 4).

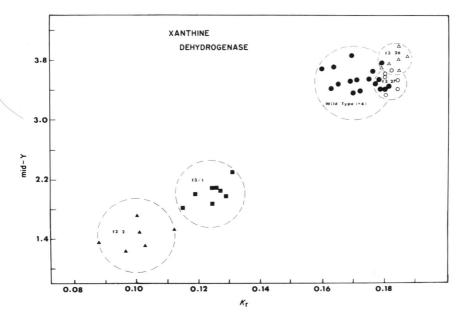

Fig. 3. Gel sieving analyses of wild type xanthine dehydrogenase (XdH), and of four different heteroallelic combinations of *ma-l* mutant alleles. In all of the stocks examined, *ma-l* and wild type, are coisogenic for chromosome 3, which contains the structural gene for *XdH*. The alleles at the *ma-l* locus were selected as XdH negatives (*ma-l* function is required for XdH activity); for analysis we have taken advantage of intraallelic complementation which occurs at the *ma-l* locus (presumably two *ma-l* mutant protein subunits combine to form a partially active multimer). The heteroallelic combinations of *ma-l* analyzed here were constructed by crossing the individual mutant alleles, so that each heteroallelic stock is heterozygous for a particular pair of complementing alleles. (From Finnerty and Johnson, 1979.)

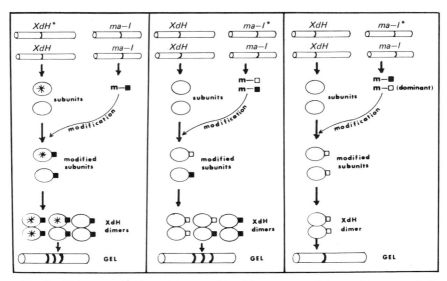

Fig. 4. Heterozygote phenotypes produced by enzyme alleles and by second-site modifiers. The gel pattern produced by heterozygosity at a modifying locus depends upon the degree of dominance of the modification. If modifier alleles lack dominance, a multiple-band gel pattern will be obtained just as expected for heterozygosity of enzyme alleles. If a modifier allele is dominant, a single gel band will be obtained. In surveys of populations containing such dominant modifiers, different single bands will be seen for each dominant allele; no heterozygotes will be detected.

The differences seen are precisely the sort reported in surveys of hidden variation (Johnson, 1977b). Thus the existing evidence does not permit unambiguous genetic interpretation of the numerous "hidden" variants reported at the *XdH* locus. They may represent alleles of *XdH*, of *ma-l*, of other loci affecting XdH levels such as *lxd* (33 on the chromosome 3) or *cin* (distal dip of the X chromosome), or any combination of these possibilities. There is no clear way to distinguish between these possibilities short of mapping the variants.

C. Functional Properties of Variants

Whether the genetic basis of the gel sieving variation is allelic or second-site modification, it appears that most if not all of it is heritable. Thus any functional differences resulting from these variants are of real genetic and evolutionary significance. The correspondence for XdH and AO between difference in shape as assessed by gel sieving and level of activity (Fig. 5) and thermal stability (Fig. 6) encourages me to expect that the

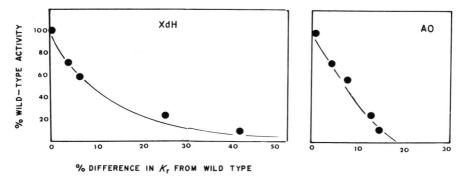

Fig. 5. Activity of conformational variants. The four variants represent four heteroallelic combinations of *ma-l* alleles, each having modified the shape (K_r) of the XdH and AO proteins to different degrees. The amount of enzyme activity is clearly related to estimates of enzyme K_r. (From Finnerty and Johnson, 1979.)

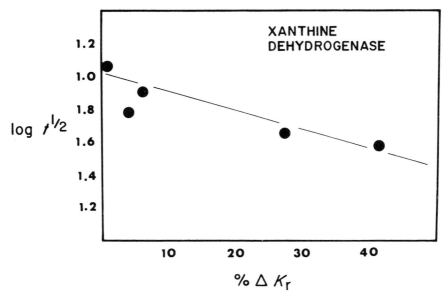

Fig. 6. Thermal stability of XdH variants. For each of the *ma-l* lines described in Fig. 3, the thermal stability of XdH was determined as $t^{1/2}$, the incubation time at 65°C required to produce a loss of 50% of XdH activity (V. Finnerty, M. McCarron, and G. Johnson, unpublished). The corresponding degrees of XdH shape modification produced by each of the *ma-l* heteroallelic combinations are determined as the percent difference in K_r from wild type, estimated from Fig. 3. Shape and stability are clearly correlated. (From Finnerty and Johnson, 1979.)

widespread differences in shape detected in my previous gel sieving surveys may reflect similar differences in function. The result of Fig. 5 is of particular importance for this reason, since it is the first data to establish that differences in K_r may indeed reflect activity differences.

Because at least a considerable portion of the hidden variation not normally detected by electrophoresis is heritable, it follows that any studies of genetic variation among loci mediating organized metabolic phenotypes will be best carried out with experimental screening procedures capable of detecting the sorts of variation described above. Posttranslational variation may play an important role in determining realized physiological phenotypes, and it is important that screening procedures be sufficiently sensitive to detect such variants. In some cases such epigenetic effects may be heritable (proteolytic effects, serial deamidation, acylation, etc.), in others, environmentally mediated. Whatever their origin, any epigenetic modification which structurally alters the enzymes under study has the potential of systematically modifying the physiological phenotype, and thus must be taken into consideration in evaluating the contribution of allelic variation to that phenotype.

D. Novel Approaches to Electrophoresis

I have recently applied to these studies two new high resolution analytic procedures. Their utility can perhaps best be appreciated by consideration of the fundamental equations which govern electrophoresis. The most basic of these says that observed mobility is a function of the diffusion constant D, the electric field strength E, and the free electrophoretic mobility of the protein,

TABLE III   Novel Approaches to Electrophoresis

Protein alteration	Procedure	Variable
Charge change	1. Change pH 2. Determine isoelectric point	Q
Shape change	1. Change buffer ion	r_i
Shape change	1. Change gel concentration 2. Determine K_r by gel sieving	η Steric
Polar $\rightleftharpoons$ nonpolar substitution	1. Change polarity of solvent 2. Change temperature	η Hydrodynamic
Subunit binding site change	1. Dilute sample	R

$$\frac{\partial C}{\partial t} = D \frac{\partial^2 C}{\partial^2 x} - \mu E \frac{\partial C}{\partial x} \tag{3}$$

The important thing is that the mobility is entirely a function of the three variables D, E, and μ. All of the analytical applications of electrophoresis are based on this relationship: If one holds E constant, then, for a defined D, the observed mobility should be a function of the protein's mobility μ. D need not, however, be kept constant; most of our progress in detecting new electrophoretic variants has come through alteration of parameters which interact to determine D

$$D = \left[\frac{Q \cdot X_1(kR)}{6\pi\eta R} \frac{(1 + kr_i)}{[1 + k(r_i + R)]} \right] \tag{4}$$

D thus is a function of net charge Q, molecular radius R, counterion radius r_i, viscosity η, and several constants. Any of these four variables are subject to experimental manipulation. When one changes the buffer pH to detect new variants, or, more precisely, determines and compares isoelectric points, one is varying Q. When one changes buffer ions to detect

TABLE IV   The Effect of D_2O on the Electrophoretic Behavior of α-Glycerophosphate Dehydrogenase in a Sample of *Colias meadii*[a]

Sample	No.	H_2O	D_2O	δ	α-GPdH[hb]
K_r	1	-0.060 ± 0.003	-0.073 ± 0.016	0.22	0.01
	2	-0.062 ± 0.009	-0.068 ± 0.016	0.10	0.04
	3	-0.067 ± 0.004	-0.075 ± 0.16	0.12	0.04
	4	-0.063 ± 0.007	-0.071 ± 0.010	0.13	0.03
	5	-0.065 ± 0.010	-0.069 ± 0.004	0.06	0.11
	6	-0.066 ± 0.010	-0.071 ± 0.011	0.08	0.03
M_0	1	1.56	2.05	0.32	0.04
	2	1.66	1.98	0.19	0.01
	3	1.77	2.23	0.26	0.01
	4	1.59	1.99	0.25	0.01
	5	1.67	1.95	0.17	0.14
	6	1.69	2.01	0.19	0.02

[a] Sixteen additional individuals were surveyed with no evidence of variation (G. Johnson, unpublished). Determinations on H_2O and on 99.8% D_2O are performed simultaneously on identical samples. δ represents the fractional difference in K_r produced by D_2O, λ the fractional difference in M_0. α-GPdH represents point to point comparison of GPdH to hemoglobin run in the same gels (Johnson, 1977a). Run to run variation is considerable unless standardized to an internal standard protein such as hemoglobin. Individual 5, normal by other criteria, shows a significant D_2O effect when standardized to hemoglobin.

new variants, one is varying r_i. When one changes acrylamide concentration in gel sieving analysis, one is varying the steric viscosity term η. The sorts of variation detected by these and other manipulations of D are outlined in Table III.

Two new procedures have been developed in an attempt to detect polar nonpolar amino acid substitutions which do not alter charge or protein shape. The first of these involves changing solvent polarity. The polarity of the solvent is readily manipulated by running gels in 99.8% D_2O. The results of one such screening procedure are presented in Table IV. One individual is clearly variant, and this individual is indistinguishable from other individuals by conventional approaches, and by both pI and gel sieving analyses.

Even greater resolution may be obtained by varying the temperature of gel sieving analysis: While D_2O increases K_r by increasing hydrophobic interactions with solvent (the strength of those interactions depending upon the polar nature of the protein, hence the detection of polarity variants), temperature decreases K_r as it interferes with hydrophobic interac-

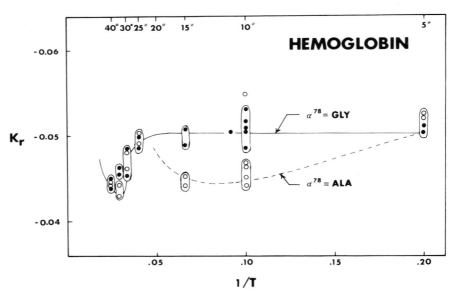

Fig. 7. High-resolution analysis of shape variation in hemoglobin. Gel sieving electrophoresis is carried out at each of eight temperatures. Two variants differing by a single amino acid substitution at position 78 (glycine for alanine) of *Peromyscus* hemoglobin α chain are presented. The two variants are indistinguishable in normal electrophoresis, but differ slightly in isoelectric focusing behavior. (From G. Johnson, unpublished.)

tions with solvent. As η is a log linear function of temperature

$$\eta = Ae^{E/RT} \qquad \text{ur} \qquad \log\eta = \frac{B}{\text{temp}} + C \qquad (5)$$

One may plot $\log K_r$ as an inverse function of temperature. Any polarity variant will be expected to differ. An example is presented in Fig. 7.

IV. PATTERNS OF VARIATION

A. Structural Pattern

It has been abundantly clear in surveys of electrophoretically detectable enzyme variation that some enzyme loci are consistently far more polymorphic than others. On the face of it, this might reflect the differing physiological roles of the enzymes (some more directly influence the physiological phenotype), or it might result from differing structural properties of the enzymes themselves. A variety of lines of investigation have concerned themselves with the second class of possibility (Harris *et al.*, 1977; Johnson, 1978). Two hypotheses of particular interest are (1) that levels of genetic variation are a function of the size of the gene and (2) that levels of genetic variation are related in some way to the propensity to form multimeric proteins.

The gene size hypothesis is most readily addressed in terms of the size of the gene product, that is to say the subunit molecular weight. One may ask whether enzymes with very large basic units tend to be more polymorphic than ones with very small ones. Several such studies have been carried out, with differing results. In an extensive study of human variation (Harris *et al.*, 1977), little difference was seen. Among 33 polymorphic loci the mean size was 45,333 daltons, while among 54 nonpolymorphic loci the mean size was 45,889 daltons. A similar lack of difference is seen in insects when molecular weight is estimated indirectly by gel sieving (Johnson, 1977d). A contrasting result, however, is reported for pooled data on a variety of species (Koehn and Eanes, 1977). Thus, the matter should probably be considered open until more data become available. This is a matter of some interest, since a random mutation interpretation of enzyme polymorphism would predict a correlation between gene size and level of variation simply because bigger genes present bigger targets for mutation. Even excepting "safe areas" required for catalysis, etc., such a general correlation ought to exist under a neutral interpretation of the variation. Because of the involvement of population size, however (effective number of alleles η_e = effective population size $N_e \times$ mutation rate μ, in the absence of selection), the exact correlation expected will depend upon the individual population chosen for analysis.

The subunit assembly hypothesis may be investigated by determining the number of subunits of each of a variety of enzymes (note that this information was required for the studies described above), and asking whether subunit number is related to degree of genetic variability. Like most issues concerning enzyme polymorphism, this question arises out of the selection/neutrality dialogue. If most electrophoretically detected variation is randomly arising and neutral in functional effect, then there ought to exist on the proteins' surface, areas critical to function where such random variation could not be tolerated, areas such as the active site or very hydrophobic "patches" involved in subunit assembly. This line of reasoning suggests that monomeric enzymes, which lack subunit binding sites, might have fewer such structural constraints, and in general tend to exhibit more variation than multimeric enzymes.

A significant relationship between subunit number and level of variability is indeed obtained in every system which has been investigated in detail. Again, Harris and co-workers' study of man provides a clear example of the sort of result which is obtained. Among 27 monomeric loci, the average heterozygosity was 0.096; among 37 dimers, average heterozygosity was significantly less, 0.071; among 19 tetramers, the average heterozygosity was lower still, 0.050. Similar relationships are reported in insects. The result is thus completely consistent with the prediction of a "neutral" viewpoint as outlined above. Please note, however, that it is also consistent with a number of alternative hypotheses.

An interesting alternative hypothesis has been suggested by Harris and co-workers, who point out that most of the polymorphism among multimeric enzyme loci is confined to those enzymes which are like monomers in that they do not form interlocus hybrid multimers (Table V). Perhaps, they point out, interlocus hybrid multimers between loci fill the physiological role that heterozygotes fill for monomers and nonhybrid formers. This is to me an exciting suggestion, as it not only provides a conceptual link

TABLE V   Polymorphism with Respect to Ability to Form Interlocus Hybrids

Enzyme-coding loci	Total No. of loci	Polymorphic No. of loci	Percentage polymorphic
Monomeric	27	15	56
Multimeric, no interlocus hybrids	38	16	42
Multimeric, forms interlocus hybrids	22	2	9

between isozymes and polymorphism (multiple loci on the one hand, and multiple alleles at one locus, on the other) but does it within a physiological framework. The implication is that the polymorphic variation which we see at some enzyme loci is but one of several alternative physiological solutions to a metabolic problem in which enzymic diversity is the key. Perhaps the presence in the same tissue of several enzymic forms with differing properties broadens the range of catalytic effectiveness or allosteric sensitivity. Any number of other conceptualizations can be devised, once within this physiological framework. Little data yet exist on the *in vitro* kinetic properties of polymorphic enzyme variants; such data when available will be of very great interest in this context.

B. Functional Patterns

Although the possibility of functional differences between electrophoretically detected variants is central to any physiological interpretation of enzyme polymorphism, relatively little direct experimental work has focused on this point. Hints about how an empirical question might be phrased arose from very insightful observations by Gillespie and Kojima (1968) and Kojima *et al.* (1970). They observed that some functional classes of enzymes seem to show far more polymorphic variation than others. Two distinctions could be noted: (1) The most variable enzymes were those whose substrates were most closely related to the immediate environment. (2) Among the remainder, enzymes involved in metabolism, those enzymes involved directly in glucose metabolism seemed the least variable. This was a particularly important observation in light of the ongoing selection/neutrality dialogue within population genetics, as levels of genetically "neutral" variation at observed enzyme loci should be random with respect to the functions of those loci, if the functions of the loci are not affected by the variation. Table VI is typical of the correlations seen between enzyme variability and involvement in the immediate external environment. Numbers of variant alleles are consistently greater for those enzymes utilizing substrates originating directly from outside the organism (starch, alcohols, peptides, etc.). The evolutionary rationale of these groupings is that enzymes utilizing internal metabolites experience far greater substrate uniformity than do enzymes whose substrates originate from the more variable external environment. A selective viewpoint is quite consistent with the finding that enzymes directly experiencing environmental diversity and variation are the more genetically variable; the hypothesis of selective neutrality, on the other hand, predicts that no consistent difference should be observed between the two functional groups.

The most important aspect of this line of rather indirect evidence is that

**TABLE VI Importance of Substrate Variability to Enzyme
Polymorphism**[a]

Enzymes	S[b]	K[c]
Utilizing internal metabolites		
Cholinesterase	2	1.00
Triose phosphate isomerase (TPI)	5	1.40
6-Phosphogluconate dehydrogenase	9	1.44
Fumarase (FUM)	8	1.50
Malate dehydrogenase (MdH)	22	1.50
α-Glycerophosphate dehydrogenase (α-GPdH)	19	1.53
Malic enzyme (ME)	16	1.62
Glucose-6-phosphate dehydrogenase (G6PdH)	6	1.83
Isocitrate dehydrogenase (IdH)	13	1.92
Phosphoglucoisomerase (PGI)	5	2.00
Aldolase	7	2.29
Hexokinase (HK)	10	2.50
Xanthine dehydrogenase	9	2.78
Phosphoglucomutase (PGM)	10	3.00
Adenylate kinase (AK)	6	3.17
Utilizing external substrates		
Alcohol dehydrogenase (AdH)	23	2.26
Amylase	2	2.50
Octanol dehydrogenase (OdH)	12	2.58
Aldehyde oxidase (AO)	15	2.73
Leucine aminopeptidase (LAP)	23	2.96
Esterases (EST)	70	3.64

[a] Values are from Johnson (1973b)
[b] S, mean number of independent samples with n 100 genomes.
[c] K, mean number of alleles observed at a frequency 0.01 in samples
with n 100 genomes.

it represents the first attempt to understand patterns of polymorphic
variation in terms of the physiological functions of the enzymes involved.
The variation/environment relationship has proved, however, somewhat
refractory to experimental analysis. The difficulty is that the role of the
environmental variation is not explicitly defined and therefore not subject
to unambiguous experimental manipulation. Very tantalizing relationships have been suggested when the immediate environment, (temperature, light, food, diversity, etc.) has been blindly manipulated (see Powell,
1971); but because the critical experimental variables are unidentified, it
has not been possible to design quantitative experiments. This would still
seem a very promising line of enquiry, if a polymorphic enzyme of know
function were examined with respect to a specific quantifiable environmental variable known to affect that function. The recent work of Schar-

loo and co-workers (1977) on variation at the *amylase* locus as affected by starch provides an example.

Enzymes not involved directly with the environment, while less variable, also exhibit considerable polymorphism. Gillespie and co-workers (Kojima *et al.*, 1970) noted that among this second class, those enzymes involved in glucose metabolism seemed less variable than the other enzymes utilizing internal metabolites as substrates. This observation has proved quite reproducible in subsequent data, but is difficult to come to grips with conceptually, as it is not immediately obvious why glucose metabolism should be genetically different from other important reaction sequences in intermediary metabolism. A way out of this blind alley is suggested by a loose correlation between polymorphism and the equilibrium constant of the reaction concerned (Johnson, 1971b): The most reversible reactions exhibited the least polymorphism. Irreversible reactions in metabolic pathways tend to be important sites of metabolic control. Perhaps polymorphism is associated with regulatory importance. This hypothesis has the virtue that the relationship between genetic polymorphism and phenotypic variation is explicit. Reactions of regulatory importance determine the physiological phenotype upon which selection acts.

The suggestion that polymorphism among enzyme loci may be related to metabolic regulation has been analyzed (Johnson, 1974). The rationale of the argument is that selection must act ultimately upon the reproductive fitness of individuals. The contributions of particular metabolic sequences to that fitness must be considered in terms of overall pathway output rather than in terms of specific reactions. Changes at loci whose enzymes regulate flow through pathways would be expected to produce far greater alterations in fitness than changes affecting enzymes which do not regulate metabolic flow. The rates of freely reversible unregulated reactions are usually so rapid that equilibrium proportions are always obtained between substrates and products of the reactions; modulation of a pathway's flux by polymorphic variation among such enzymes seems unlikely because activities would have to be reduced by several orders of magnitude.

So, in practice, which are the reactions which regulate the rates of metabolic processes? These may be identified by equilibrium perturbation experiments. When a metabolic sequence is operating at steady state *in vivo*, each reaction proceeds at an identical rate. Thus, under steady-state conditions most enzymes of a pathway are working well below their maximum capacity, while one or a few enzymes dictate the rate of flux through the pathway. To identify the steps regulating flow, one may vary the flux through the pathway, monitoring the concentrations of intermediates. The ratios of substrate to product for all uncontrolled reactions are independent of the rate of overall flux through the pathway. Rate-limiting or controlled reactions, however, will show a deviation from the equilibrium ratio of product to substrate. Thus, when the observed ratio of substrate to product deviates significantly *in vivo* from the ratio predicted by its thermodynamic equilibrium, this can be taken as *prima facie* evidence that the reaction of this enzyme is of regulatory importance in the overall pathway.

Estimates of metabolite concentrations obtained from perturbed pathways *in vivo* are not certain indicators of regulatory involvement, because intracellular compartmentalization may produce misleading results. Such estimates do provide, however, the best general characterization of metabolic regulatory involvement currently available. A wide variety of enzymes in many organisms and tissue systems have been studied in this way. Some of them for which genetic variation studies have also been carried out are listed in Table VII. Enzymes are assigned to categories according to the values reported for substrate–product ratios measured *in vivo* after perturbation, relative to expected equilibrium values. When the perturbed substrate–product ratios did not deviate significantly (by greater than one order of magnitude) from equilibrium, the reaction is characterized as "nonregulatory."

A comparison of levels of polymorphism between regulatory and nonregulatory enzyme loci is presented in Table VIII. Nonregulatory reactions are clearly less polymorphic. Subsequent analyses utilizing more extensive data sets have confirmed the same general relationship (Powell, 1975). The result is of real importance, as it, like the subunit assembly work of Harris and co-workers (Harris *et al.*, 1977), suggests that the polymorphic variation which we see in intermediary metabolism is a physiological solution to a metabolic problem in which enzymic diversity is the key. Here an explicit experimental approach is quite practical, as the levels of metabolite intermediates may be monitored under perturbation for each of several alternative genotypes of polymorphic enzyme loci. The experimental methodology is available; an example is the elegant series of procedures developed by Sacktor (1970) in studying metabolite fluxes during the initiation of insect flight. I regard the potential of this approach as very promising, although little has been accomplished with it to date.

There are difficulties in a too casual interpretation of Table VIII: (1) Pleiotropism makes simple assignments difficult. Alcohol dehydrogenase, for example, may function in lipid metabolism, vitamin A synthesis, juvenile hormone synthesis (in *Drosophila*), and vision in the forward direction (aldehyde to alcohol), as well as possibly playing a role in the breakdown of dietary alcohols in the reverse direction (alcohol to alde-

TABLE VII The Regulatory Role of Some of the Enzymes Commonly Examined for Polymorphism[a]

Enzyme	Evidence
Regulatory enzymes	
Adenosine deaminase	Sigmoidal kinetics *in vitro*
Adenylate kinase (AK)	Substrate-product ratio significantly displaced from equilibrium
Glucose-6-phosphate dehydrogenase (G6PdH)	Sigmoidal kinetics; substrate–product ratio significantly displaced from equilibrium
Glyceraldehyde-3-phosphate dehydrogenase	Substrate–product ratio significantly displaced from equilibrium
Hexokinase (HK)	Substrate–product ratio significantly displaced from equilibrium
Isocitrate dehydrogenase (IdH)	Sigmoidal kinetics; substrate–product ratio may be significantly displaced from equilibrium
Phosphoglucoisomerase (PGI)	Substrate–product ratio sometimes displaced from equilibrium
Phosphoglucomutase (PGM)	Substrate–product ratio significantly displaced from equilibrium
Phosphoglycerate Kinase	Substrate–product ratio significantly displaced from equilibrium
Phosphofructokinase (PFK)	Substrate–product ratio significantly displaced from equilibrium
Pyruvate kinase (PK)	Substrate–product ratio significantly displaced from equilibrium
Nonregulatory enzymes	
Aldolase	Substrate–product ratio at equilibrium; isotope randomization indicates reaction is very fast
Carbonic anhydrase	Reaction kinetics very fast
Isocitrate dehydrogenase (IdH)	Substrate–product ratio at equilibrium
Lactate dehydrogenase (LdH)	Substrate–product ratio at equilibrium
Malate dehydrogenase (MdH)	Substrate–product ratio at equilibrium
6-phosphogluconate dehydrogenase	Substrate–product ratio at equilibrium

[a] Citations containing experimental data may be obtained from Johnson, 1974.

hyde). (2) Several enzymes are represented by multiple loci, some of which are regulatory *in vivo* and others not. It is difficult to know *a priori* which is which.

The fundamental weakness with this approach, attempting to assess the functional significance of patterns of enzyme polymorphism in terms of regulatory involvement, is that it describes the physiological phenotype

TABLE VIII   Metabolic Patterns of Polymorphism[a]

	Drosophila		Small vertebrates		Man	
	No. loci	HET	No. loci	HET	No. loci	HET
Regulatory	10	0.19	8	0.14	21	0.13
Nonregulatory	7	0.06	9	0.06	23	0.005

[a] Data are expressed as average heterozygosities, HET. Values are from Johnson, 1974.

which polymorphism produces in too simple a manner, attempting to index it in terms of a single regulatory reaction. In reality, the situation is more complex, with the interaction of many factors serving to define metabolic states. However, the utility of the regulatory involvement approach is that it serves to identify a relationship between enzyme polymorphism, on the one hand, and physiological aspects of the phenotype, on the other. It points the way to an empirical approach to the fundamental question of what that relationship might be. If we are going to study the phenotypic consequences of enzyme polymorphism in an attempt to assess selective significance, then we are going to have to look at *physiological phenotypes*. That a primitive index such as given in Table VII produces such an interesting result as given in Table VIII suggests that such a look may be very fruitful.

V. COORDINATE METABOLIC PHENOTYPES

A. Single Locus Studies

The simplest solution to the quandary posed by complex physiological phenotypes is, of course, to examine loci whose enzymes do not have complex physiological expression, but rather produce simple defined effects upon the phenotype. The rationale of such single-locus approaches has been summarized (Clarke, 1975; Johnson, 1975a). The clearest examples of such single gene polymorphisms are the alcohol dehydrogenase polymorphism in *Drosophila,* an apparent adaptation to toxic levels of alcohol in fermenting fruit (Day *et al.,* 1974); lactate dehydrogenase polymorphism in fish, an adaptation of physiological redox levels to changes in temperature (Merritt, 1972); α-glycerophosphate dehydrogenase polymorphism in insects, an adaptation of flight muscle redox levels to changes in temperature (Johnson, 1976a); glucose-6-phosphate dehydrogenase and hemoglobin polymorphisms in man, which confer resistances to malarial parasites (Ingram, 1957); and recently leucine aminopeptidase (LAP) in marine mussels, an apparent adaptation of ion balance to salinity stress (R. Koehn, personal communication). In all of these cases, polymorphic variation represents an apparently direct adaptation to changes in a specific environmental factor. The relationship is in each case a direct cause and effect one, and may in principle be fully understood in terms of the difference in that enzyme's function which the polymorphism produces.

Single-locus approaches are basically unsatisfactory for two interrelated reasons. (1) The selection coefficient of an allele at one locus may be affected by the genotype of another locus. (2) The recombination fraction between the two loci is subject to change. The point is that the target of selection is the operational phenotype, whether it be flower morphology or respiratory efficiency, and many loci interact to determine such phenotypes. In addition, because selection may involve multiple loci, the degree of linkage and thus of coordination between the loci is itself subject to evolutionary modification. Any single-locus study ignores such interactions and carries out one-locus–environment comparison in an attempt to understand adaptation may misinterpret or even fail to detect key elements in the process.

B. The Metabolic Phenotype

For many of the loci reported to be polymorphic in natural populations, simple and direct relationships with the environment seem unlikely. Many of these enzymes are intimately involved in intermediary metabolism, and a change in the activity of one may influence the functioning of many others. Thus, a change in hexokinase, which generates glucose 6-phosphate, cannot help but affect the reactions of phosphoglucomutase, phosphoglucoisomerase and glucose-6-phosphate dehydrogenase, all of which use glucose 6-phosphate as a substrate (Johnson, 1976c). The highly coordinated nature of intermediary metabolism suggests that if polymorphic alleles at regulatory enzyme loci are functionally different, then the particular allele present at one locus of a rate-determining enzyme will importantly affect the activity not only of its own pathway but also of many other reactions. If, for a network of related regulatory enzymes, each locus maintains that allele which provides optimal activity in a particular habitat, then one may speak of a metabolic phenotype adapted to that habitat. It is important to realize that because the particular functional variant occurring at each regulatory locus influences the overall physiological state of the individual, selection on metabolic phenotypes implies

selection on allozymic genotypes. To understand the process, the genotypes of each of the loci importantly contributing to the phenotype must
be known.

To test the hypothesis that polymorphism among the loci of metabolic
enzymes reflects selection for alternative integrated metabolic phenotypes, it is very desirable that investigations focus simultaneously on several different sorts of experimental information.

1. The loci examined must adequately characterize the metabolic
processes defining the physiological phenotype under study. Both regulatory and nonregulatory loci should be included in the analysis. Indeed, as
complete as possible a collection of all the enzymes involved in the metabolic processes is desirable, although practical considerations may make
this difficult. It may also be useful to include within the analysis several
loci which are known not to be involved in the postulated metabolic
phenotype, in order to see that they vary independently.

2. A genetic screening procedure must be used which is sensitive
enough to detect all of the variants. This is important, because spurious
associations may be suggested by insensitive techniques and because
other significant associations may be hidden by them. Recent studies have
shown that when different alleles are combined together into electrophoretic classes, gene associations (linkage disequilibrium) are often dis-
guised, so that there is a much lower probability of detecting associations
which do occur (Zouros *et al.*, 1977). Thus, in attempting to assess the
degree to which genic variation at enzyme loci acts to modulate physiological processes, we need not only to phrase our questions in terms of
loci which adequately characterize the process under study but also to
employ screening procedures which will detect genic variation affecting
that process.

3. Genetic analysis must be carried out to ensure that all electrophoretically detected variants are allelic and not produced by some posttranslational process. In order to assign an explicit multilocus genotype to each
individual, it is critical that the selected array of loci all be examined in
each individual.

4. The environmental parameters which significantly affect the physiological phenotype under study must be known and quantifiable. In particular, the natural microhabitats of the populations under study must be
characterized for the presumptively selective environmental variable.
Several populations should be studied whose microhabitats differ in this
regard. It is important that the sample represent resident individuals (not
migrant individuals from adjacent populations with potentially different
microhabitats) and that the particular populations selected for study not
be themselves spatially subdivided environmentally or genetically.

5. An *in vitro* kinetic analysis of each enzyme variant should be carried out to see if any functional difference is indicated. If the selected environmental variable may be manipulated (temperature, pH, ionic strength, inhibitor concentrations, etc.), kinetic comparisons should be carried out over a range of conditions consistent with the microhabitat characterization.

6. When kinetic comparisons of polymorphic variants at particular reactions suggest the possibility of physiological effect, an *in vivo* physiological analysis should be carried out to see if the allelic difference at that locus actually works to produce the postulated physiological effect. Again, the appropriate range of environmental conditions must be considered. Analyses must be carried out on individuals of known genotype, so that other loci critical to the physiological phenotype are known not to vary.

7. Where *in vivo* comparisons of allelic variants indicate real modification of the physiological phenotype, demographic life history analyses of the sampled populations should be carried out, in order to ascertain whether the genetically induced difference in phenotype actually results in differential reproductive fitness.

C. Two Promising Systems for Study

1. Glycolysis and Anaerobic Metabolism

The principal enzyme reactions of glycolysis (and its reverse, gluconeogenesis) are well known. For a variety of tissues, relatively complete and simultaneous analyses of all the intermediates and cosubstrates have been carried out. The results indicate uniformly that the regulation of glycolytic flow is at the early and late irreversible reactions phosphofructokinase (PFK) and pyruvate kinase (PK). Gluconeogenesis may occur in a separate intracellular region, independently regulated at the reaction fructose-1,6-diphosphatase (FDPase). These three reactions, PFK, PK, and FDPase thus comprise central points of the metabolic control of anaerobic metabolism. Simultaneous survey of polymorphic variation at a variety of glycolytic enzyme loci, including both these three and others known not to be points of regulation (*F6P, G6P, Triose P, 3PGA, Pep*), will provide multilocus genotypes which permit an unbiased determination of the degree to which enzyme polymorphism is associated with the phenotype-determining steps and the degree to which these key enzyme loci are coordinate in their genotypic expression. *In vivo* monitoring of metabolite levels is practical, as discussed earlier.

The degree of anaerobic metabolism may be tightly coupled to environmental constraints, particularly in plants subjected to periodic flooding.

Under prolonged flooding, roots may be subjected to anaerobic conditions for significant periods, leading to potentially harmful accumulations of ethanol. Long-term adaptation to "flood tolerance" has been reported to involve loss of activity of NADP-malic enzyme, and use of malate dehydrogenase as an alternative mechanism of restoring the NADH–NAD balance, resulting in the accumulation of malate, which is not toxic. In a "flood-intolerant" plant, malate is converted to pyruvate by NADP-malic enzyme, which is readily oxidized to acetaldehyde and then to ethanol by the sequential action of pyruvate decarboxylase and alcohol dehydrogenase (AdH). An alternative proposal is that adaptation to flooding involves changes at the *pyruvate decarboxylase* locus itself, altering production of acetaldehyde. As high levels of acetaldehyde and ethanol are toxic, it is easy to see why under anaerobic conditions malate or pyruvate accumulation is preferable.

The enzyme activity of AdH may be monitored *in vivo* under various levels of moisture stress by using an intact tissue assay (R. Mitra and J. Varner, personal communication). The *in vivo* assay depends on the loss to water of tritium from specifically tritiated substrates. It will be necessary to check malate dehydrogenase, malic enzyme, and pyruvate decarboxylase as alternative modes of response, particularly in cases of increased water potential; for each of these reactions, *in vivo* assays are potentially realizable.

The interesting evolutionary question in the context of this article concerns plant populations living in habitats where they are subjected to the moisture stress of periodic flooding. In plants exposed to less chronic stress, modulation of alcohol dehydrogenase activity at low oxygen tension may be extremely important. Simultaneous survey of polymorphic variation at the seven enzyme loci *PEP-Carbox, PK, MdH, NADP-ME, pyruvate decarboxylase, AO,* and *AdH* will permit an unbiased determination of the degree to which enzyme polymorphism is associated with moisture tolerance, and in particular with the phenotype-determining reaction catalyzed by NADP-malic enzyme.

It is important not to lose track of the fact that other loci may also be involved in a response to moisture stress. Certain peroxidases catalyze cell wall lignification. Their levels change with developmental stage and with conditions of stress. Catalase also appears to have an obvious function in responding to oxygen stress. Its level of activity and localization change with developmental stage, and its level is high in the microbodies of tissues in which lipid reserves are rapidly broken down. Environmental influence on lipid metabolism should be reflected in catalase levels, although the mode of this interaction is open to some discussion.

In vivo assays of the combined activity of these enzymes in the root are practical, as H_2O_2 freely diffuses out of roots; a root submerged in dilute H_2O_2 and monitored with an O_2 electrode ought to provide a direct approach. Again, it should be noted that the assay in no way injures the plant.

2. Photosynthetic Carbon Fixation

Photosynthesis is one of the most biologically important, metabolically unique, and intensively studied physiological processes in plants. Although photosynthetic carbon fixation is well suited to the type of approach suggested here, little data on genic variation at photosynthetic loci among natural populations of higher plants are available, although studies have been initiated in my laboratory (Enama, 1977). Figure 8 illustrates

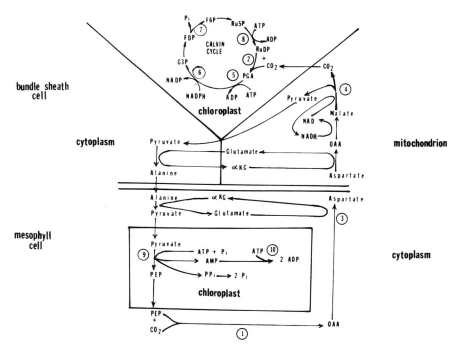

Fig. 8. Principal reactions of photosynthetic carbon fixation in a C_4 NAD-malic enzyme plant. The numbered enzymes are (1) Phosphoenolpyruvate carboxylase (EC 4.1.1.38); (2) ribulose-1,5-diphosphate carboxylase (EC 4.1.1.39); (3) glutamate-oxalacetate transaminase (EC 2.6.1.1); (4) NAD-malic enzyme (EC 1.1.1.39); (5) phosphoglyceric acid kinase (EC 2.7.2.3); (6) NADP-glyceraldehyde-3-phosphate dehydrogenase (EC 1.2.1.11); (7) fructose-1,6-diphosphatase (EC 3.1.3.11); (8) phosphoribulokinase (EC 2.7.9.1); (9) Pyruvate phosphate dikinase (EC 2.7.9.1); (10) adenylate kinase (EC 2.7.4.3).

the principle reactions of C_4 photosynthesis in a NAD-malic enzyme plant. Regulatory aspects of photosynthetic carbon metabolism have recently been extensively reviewed (Kelly *et al.*, 1976). The principal sites of regulation in C_4 photosynthesis are the reactions catalyzed by PEP carboxylase, RudP carboxylase, and fructose diphosphatase.

An in depth survey of enzyme polymorphism among the reactions of Fig. 8 would be of major value to population genetics. If care is taken to characterize all loci in each individual, the resulting multilocus genotypes will permit a detailed analysis of the genetic basis of the photosynthetic phenotype. Genic interaction between rate-determining loci, if it occurs, will be clearly characterized by such a study.

Most importantly, the rate of photosynthetic carbon fixation may be monitored *in vivo* under various temperature, moisture, and light regimes without harming the living plant (see, for example, Berry, 1975; Bjorkman *et al.*, 1972).

D. Epistasis

In considering metabolic phenotypes we have discussed reaction sequences which act in a concerted way to determine the rate of a particular metabolic process, such as glycolysis or photosynthetic carbon fixation. In a real physiological sense this analysis remains an incomplete one, as it does not take into account interacting metabolic systems. The metabolic flux through glycolysis has important influences upon pentose metabolism and aerobic respiration. Indeed, most major metabolic sequences are interrelated by a complex web of regulatory controls, mediated by allosteric interactions with cofactors, such as NAD, NADP, and ATP, or effectors, such as cyclic AMP. Without this coordination, an integrated metabolism would be impossible, and yet it poses an evolutionary problem of the first order (Johnson, 1975b): different pathways almost always have different Q_{10}s with respect to habitat variables. How is the integration to be maintained in a variable environment?

An example of epistatic interactions is provided by genic variation in binding affinities for NAD^+. Cellular redox levels (such as the ratio of NAD^+ to NADH) are set by loci such as *LdH* or $\alpha GPdH$, which serve to regenerate NAD^+ at a variable level relative to NADH concentrations. To the extent that the functioning of such enzymes is sensitive to environmental parameters, such as temperature, modulation of their activity has widespread consequences through the currency of NAD^+ allosteric binding. Genetic polymorphism at a cellular redox-determining locus, as a phenotypic response to environmental variance, expresses itself by acting on a range of loci sensitive to NAD^+. The effect may thus be pleotropic.

The key evolutionary question concerns the influence of metabolic integration on the strength of selection in a heterogeneous environment: In what sense are levels of metabolic integration a function of environmental heterogeneity?

VI. CONCLUSION

The experimental approaches suggested in this chapter involve considerable empirical effort. In individual cases, it will usually prove very difficult to obtain all the experimental information which an ideal analysis might require, and few investigators will choose to pursue all lines of investigation. The physiological rationale presented here does, however, provide guidelines which can help to focus various lines of experimentation onto common areas, and so shape our investigation of genetic polymorphisms among enzyme loci. Several points need to be kept clearly in mind.

1. Variants which are detected by highly sensitive experimental procedures may often prove nonallelic, reflecting posttranslational modifications (Watt, 1977). It is very important that any investigation into the genetic basis of enzyme polymorphic variation adequately address this point and provide a clear empirical demonstration of allelism in those cases where genetic inference is made. It is equally important that epigenetic nonallelic variants, when detected, be properly investigated, as they may make significant contributions to the realized phenotypes. Indeed, such variation may itself be under genic control.

2. Adaptive responses importantly influencing physiological phenotypes need not always involve exclusively the structural genes. Important changes can be brought about by changes in genetic regulatory elements (McDonald *et al.*, 1977; Edwards *et al.*, 1977). The possibility of polymorphism among such genetic control regions is one of the most exciting questions currently being pursued in population genetics.

REFERENCES

Bernstein, S., Throckmorton, L., and Hubby, J. (1973). *Proc. Natl. Acad. Sci. U. S. A.* **70,** 3928–3931.
Berry, J. (1975). *Science* **188,** 644–650.
Bjorkman, O., Nobs, M., Mooney, H., Troughton, J., Nicholson, F., and Ward, W. (1972). *Carnegie Inst. Washington, Yearb.* **71,** 748–767.
Chovnick, A., Gelbart, W., and McCarron, M. (1977). *Cell* **11,** 1–10.
Clarke, B. (1975). *Genetics* **79,** 101–113.

Cochrane, B. (1976). *Genetics* **83**, s16.
Coyne, J. (1976). *Genetics* **84**, 593–607.
Day, T., Hiller, P., and Clarke, B. (1974). *Biochem. Genet.* **11**, 155.
Edwards, T., Candido, E., and Chounick, A. (1977). *Mol. Gen. Genet.* **154**, 1–6.
Enama, M. (1977). *Carnegie Inst. Washington, Yearb.* **75**, 409–410.
Ewens, W. (1977). *Adv. Hum. Genet.* **8**, 67–134.
Ferguson, K. (1964). *Metab. Clin. Exp.* **13**, 985–1002.
Finnerty, V. (1976). *In* "The Genetics and Biology of Drosophila" (M. Ashburner and E. Novitski, eds.), Vol. 1B, pp. 721–760. Academic Press, New York.
Finnerty, V., and Johnson, G. (1979). *Genetics* **91**, 695–722.
Finnerty, V., McCarron, M., and Johnson, G. (1979). *Mol. Gen. Genet.* **172**, 37–43.
Gillespie, J., and Kojima, K. (1968). *Proc. Natl. Acad. Sci. U. S. A.* **61**, 582–585.
Hamrick, J. (1978). *In* "The Population Biology of Plants" (O. Solbrig *et al.*, eds.), Columbia Univ. Press, New York (in press).
Harris, H. (1966). *Proc. R. Soc. London, Ser. B* **164**, 298–310.
Harris, H. (1971). *J. Med. Genet.* **8**, 444–452.
Harris, H., Hopkinson, D., and Edwards, Y. (1977). *Proc. Natl. Acad. U. S. A.* **74**, 698–701.
Ingram, V. (1957). *Nature (London)* **180**, 326–328.
Johnson, G. (1971a). *Proc. Natl. Acad. Sci. U. S. A.* **68**, 997–1001.
Johnson, G. (1971b). *Nature (London)* **232**, 347–348.
Johnson, G. (1973a). *Annu. Rev. Ecol. Syst.* **4**, 93–116.
Johnson, G. (1973b). *Nature (London), New Biol.* **243**, 151–153.
Johnson, G. (1974). *Science* **184**, 28–37.
Johnson, G. (1975a). *Stadler Genet. Symp.* **7**, 91–116.
Johnson, G. (1975b). *In* "Molecular Evolution" (F. Ayala, ed.), pp. 46–59. Sinauer Assoc., Sunderland, Massachusetts.
Johnson, G. (1976a). *Biochem. Genet.* **14**, 403–426.
Johnson, G. (1976b). *Genetics* **83**, 149–167.
Johnson, G. (1976c). *Ann. Mo. Bot. Gard.* **63**, 248–261.
Johnson, G. (1977a). *Annu. Rev. Ecol. and Syst.* **8**, 309–328.
Johnson, G. (1977b). *Biochem. Genet.* **15**, 665–693.
Johnson, G. (1977c). *Genetics* **87**, 139–157.
Johnson, G. (1977d). *In* "Isozymes" (M. C. Rattazzi, J. G. Scandalios, and G. S. Whitt, eds.), Vol. 2, pp. 1–21. Alan R. Liss, Inc., New York.
Kelly, G., Latzo, E., and Gibbs, M. (1976). *Annu. Rev. Plant Physiol.* **27**, 181–205.
Koehn, R., and Eanes, W. (1977). *Theor. Popul. Biol.* **11**, 330–341.
Kojima, K., Gillespie, J., and Tobari, Y. (1970). *Biochem. Genet.* **4**, 627–637.
Lewontin, R. C. (1974). "The Genetic Basis of Evolutionary Change." Columbia Univ. Press, New York.
Lewontin, R., and Hubby, J. (1966). *Genetics* **54**, 595–609.
McDonald, J., Chamber, G., David, J., and Ayala, F. (1977). *Proc. Natl. Acad. Sci. U. S. A.* **74**, 4562–4566.
Milkman, R. (1976). *Biochem. Genet.* **14**, 383–387.
Merritt, R. (1972). *Am. Nat.* **106**, 173–184.
Petrakis, P., and Brown, C. (1970). *Comp. Biochem. Physiol.* **32**, 475–487.
Powell, J. (1971). *Science* **174**, 1035–1036.
Powell, J. (1975). *Evol. Biol.* **8**, 79–119.
Sacktor, B. (1970). *Adv. Insect Physiol.* **7**, 267–347.
Sampsell, B. (1977). *Biochem. Genet.* **15**, 971–988.

Scharloo, W., Van Dijken, F., Hoorn, A., de Jong, G., and Thorig, G. (1977). *In* "Measuring Selection in Natural Populations" (F. Christiansen and T. Fenchel, eds.), pp. 131–148. Springer-Verlag, Berlin and New York.

Singh, R., Hubby, J., and Lewontin, R. (1974). *Proc. Natl. Acad. Sci. U. S. A.* **71,** 1808.

Singh, R., Hubby, J., and Throckmorton, L. (1975). *Genetics* **80,** 637–650.

Singh, R., Lewontin, R., and Felton, A. (1976). *Genetics* **84,** 609–629.

Thorig, G., Schoone, A., and Scharloo, W. (1975). *Biochem. Genet.* **13,** 721–730.

Trippa, G., Loverre, A., and Gatamo, A. (1976). *Nature (London)* **260,** 42–44.

Watt, W. (1977). *Genetics* **87,** 177–194.

Wright, T., and MacIntyre, R. (1965). *J. Elisha Mitchell Sci. Soc.* **81,** 17–19.

Zouros, E., Golding, G., and MacKay, T. (1977). *Genetics* **85,** 543–550.

Index